D0789033

The Life Game

also by Nigel Calder:

Violent Universe
The Mind of Man
Restless Earth

Nigel Calder

The Life Game

EVOLUTION AND THE NEW BIOLOGY

THE VIKING PRESS NEW YORK

TO PHILIP DALY

Published in 1974 by The Viking Press, Inc.
625 Madison Avenue, New York, N.Y. 10022
SBN 670-42798-5
Library of Congress catalog card number: 73-19041
Text printed in U.S.A.
Color section printed in England

Acknowledgment is made to Random House, Inc., for the quotation from "Culture," from *The Collected Poetry of W. H. Auden*. Copyright 1945 by W. H. Auden.

Contents

Acknowledgments

Acknowledgment is due to the following for their permission to reproduce illustrations in this book. Figures in **bold** indicate colour plates facing.

David Attenborough, 116; Professor Elso S. Barghoorn, **72**, 85; Professor George W. Barlow, 52; Roland T. Bird, 108; Black Star, 39; British Museum, Natural History, 21; Buffalo Museum of Science, 111; Dr F. M. Carpenter, 100, **113**; Professor Hampton Carson, 32; M. S. Carson, 28; Bruce Coleman Ltd, **48**; Gene Cox, **88**, **89**; CSIRO, 107; Philip Daly, **41**, 69; Dr J. T. Finch, 10; Michael Freeman, **16** (courtesy Dr John Baker), **32** (courtesy British Museum, Natural History), 51, 73, 77, 96, **96**, 128 (courtesy British Museum, Natural History); John Goodyer, 79; Dr A. V. Grimstone, 88, 89; C. A. W. Guggisberg, Bruce Coleman Ltd, 127; Dr L. B. Halstead, 122; Robert Harding Associates, **40-1**; Dr Michael Harris, Bruce Coleman Ltd, **33**, **40**, **65**; Stuart Harris, 62; Eric Hosking, 36; Imitor, 115; Dr D. Jackson, 132; Kenneth Y. Kaneshiro, 31; Keystone Press Agency, 13; Dr A. Kortlandt and the Sixth Netherlands Chimpanzee Expedition, University of Amsterdam, 118; Karoly Kutasi, 113; Dr S. M. Lewis, 8; Dr A. R. Lieberman, 86; Dr O. L. Miller, Jr (from 'The Visualization of Genes in Action', *Scientific American*, March 1973), 27; Museum of Mankind (photos Derrick E. Witty), 21, 22; L. Margulis, 80; NASA, 74; Orion Press, **64**; Axel Poignant, **17**, 21; John Perkins, 122, 123; Kevin Rowley, 78; Royal Anthropological Institute Library (photos Derrick E. Witty), 22; Dr H. Chica Schaller, 93, 94; Dr Dennis Summerbell, **97**; US National Museum, **112**; Professor E. O. Wilson, **49**.

Diagrams and maps by Richard Bonson.

Author's note

This book is the fraternal twin of a major television programme of the same title. Although separately conceived, it has drawn on the same fund of information and advice, generously provided by more than a hundred biologists in many parts of the world. Regrettably, I have not been able to repay my debt to them by mentioning all their names or dealing at adequate length with their individual lines of research. Special thanks go to Manfred Eigen, Motoo Kimura, Richard Lewontin, John Maynard Smith and Lewis Wolpert; they helped to shape my approach to the subject without, of course, being responsible for any shortcomings in the book. I am grateful, too, to the BBC and its partners in the television production, for the opportunity to travel around the world.

The television programme, *The Life Game*, was first transmitted on BBC2 on 13 October 1973. It was made by the BBC as a coproduction with WNET (New York), Sveriges Radio (Stockholm), ABC (Sydney), Studio Hamburg, KRO (Hilversum), BRT (Brussels) and OECA (Toronto). The executive producer was Adrian Malone, assisted by Stuart Harris, who was also the studio director. Eugene Carr and Keith Hopper were the principal film cameramen and the film editor was James Latham. Graphics were by Charles McGhie, visual effects by Michealjohn Harris and the studio designer was Stewart Marshall. The programme was written by Nigel Calder.

In the US, the television programme is entitled *Genetics and the Life Game*.

Further reading

In giving a fresh account of evolution in the light of modern ideas I have passed lightly over many aspects of the subject that are commonly regarded as indispensable. The following books, among many others, will help the interested reader to dig more deeply. They are listed in approximate order of increasing difficulty for the non-biologist.

The Origin of Species by Charles Darwin (one modern edition is Penguin; Harmondsworth 1968)

Nature edited by James Fisher and others (Macdonald; London 1960)

The Science of Genetics by Charlotte Auerbach (Hutchinson; London 1969)

The Theory of Evolution by John Maynard Smith (Penguin; Harmondsworth 1966)

On Evolution by John Maynard Smith (Edinburgh University Press; Edinburgh 1972)

Evolution, Genetics and Man by Theodosius Dobzhansky (Wiley; New York 1955)

A Century of Darwin edited by S. A. Barnett (Heinemann; London 1958)

The Roots of Mankind by John Napier (Allen and Unwin; London 1971)

Populations, Species and Evolution by Ernst Mayr (Belknap/Harvard; Cambridge, Mass. 1970)

The Origins of Life by L. E. Orgel (Wiley; New York 1973)

Origin of Eukaryotic Cells by Lynn Margulis (Yale University Press; New Haven 1970)

Molecular Biology of the Gene by James D. Watson (Benjamin; New York 1970)

Chemical Evolution by Melvin Calvin (Oxford University Press; London 1969)

The Genetics of Human Populations by L. L. Cavalli-Sforza and W. F. Bodmer (Freeman; San Francisco 1971)

Theoretical Aspects of Population Genetics by Motoo Kimura and Tomoko Ohta (Princeton University Press; Princeton 1971)

Deformed red blood cells from a human being with the sickle-cell gene. The grim disease of sickle-cell anaemia represents evolution in progress among humans today and is the prime case of a genetic condition now understood in chemical detail.

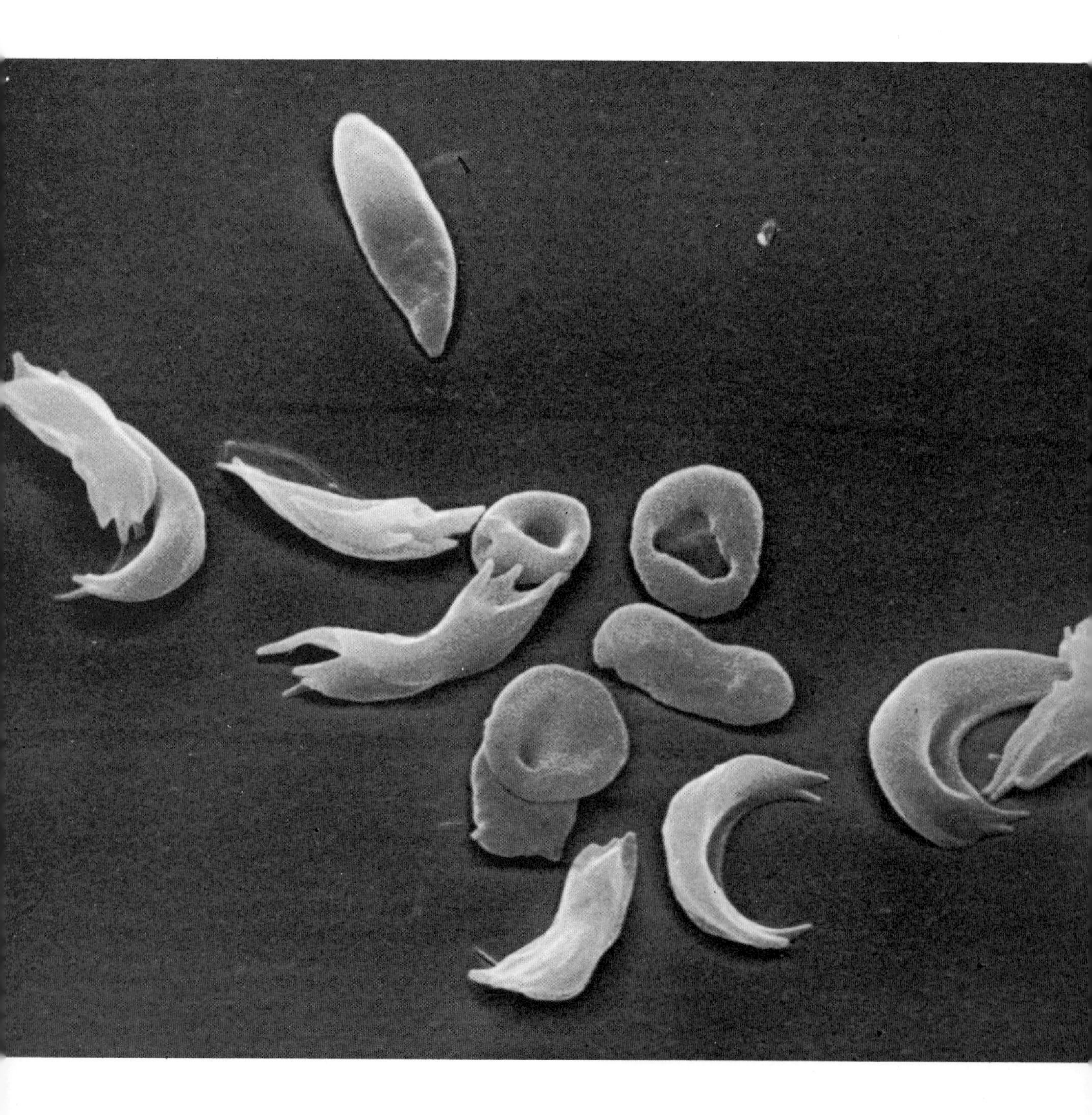

Chapter I Messages from the Dead

What should we tell Charles Darwin, if he rejoined us after ninety-odd years in the grave? He would need to know about the survival of the quickest, when it came to crossing the street. And he would be glad, of course, that men and women still interest themselves in that question of questions – how do we come to be here? Learning that the state of California is now editing its school biology books so as to avoid offence to the Bible thumpers might irritate him, but probably no more than overhearing the remark, 'Darwin said it all'. He knew better than anyone what gaps he and his contemporaries left in the account of evolution. Indeed, biologists have been finding out so much that is new, since Darwin's time, that it would be hard to know where to begin.

Above all, Darwin would want to find the scientific action around the world, in the laboratories and in the wild, and to meet the biologists who are making the big discoveries today. My own good fortune has been the opportunity to make such a journey. My first task was to script a television programme for the BBC and coproducing organisations from seven other countries; the second aim was the book that follows. In my travels I was always accompanied, in my thoughts, by the young naturalist who made the journey the slow way in HMS *Beagle*.

He would have been fascinated by new evidence that the hereditary differences between individuals – the key to evolution – are far richer than biologists ever imagined. To one who stressed the importance of geographical barriers, for allowing new species of plants and animals to evolve, the movements of whole continents, newly confirmed, would be a dazzling illumination of the history of life. And the unimaginable length of time that Darwin knew must be available for the course of evolution on Earth – how that has stretched since his day! Radioactive dating techniques now put the age of the Earth at 4600 million years, and the oldest known fossils, of microbes, are 3200 million years old. As for ourselves, the latest finds show that human predecessors living in Kenya three million years ago had already attained fully half our brain capacity.

But Darwin would be just as baffled as any member of the public today by the jargon of the new biology. Worse off, perhaps; never mind that for him the *Journal of Molecular Biology* might as well be written in Sanskrit, Darwin did not even know about the genes. He wrote:

> The laws governing inheritance are quite unknown; no one can say why the same peculiarity in different individuals of the same species, and in individuals of different species, is sometimes inherited and sometimes not so; why the child often reverts in certain characters to its grandfather or grandmother or other much more remote ancestor.

Our great advantage over Darwin is that these mysteries are wholly dispelled. We not only know that 'inheritance' comes from our ancestors as a mass of different messages called genes; we can even say in what chemical language these messages from the dead to the living are written. At the centre of current efforts in evolutionary science is the restatement of Darwin's great theme in the light of these and related discoveries.

For bringing Darwin swiftly up to date, one recent subject of research serves best. The topic, sickle-cell anaemia, is a painful one – literally and often fatally painful for the black babies who suffer from this hereditary disease of the blood. Even so, sickle-cell anaemia is the prime example of continuing human evolution and of its new interpretation in the chemistry of the genes.

What goes disastrously wrong in those who suffer from sickle-cell anaemia is the machinery for carrying oxygen from the lungs, around the body, to all

What goes wrong in sickle cells. In the absence of oxygen, molecules of sickle haemoglobin stick together and form these long tubes revealed by the electron microscope. They deform the cells and prevent them from working properly.

the tissues that need it. The little, round, red blood cells that have this task become distorted, often into the shape of a sickle. It happens rather easily and eventually fatally in sufferers from the disease. In their relatives who are 'carriers' of the disease, sickling occurs only in quite exceptional circumstances of breathlessness; otherwise they lead normal lives, even as sportsmen. In the United States recently, public comment and even legislation have confused these two very different conditions. Political and medical fecklessness in the Federal Government's current sickle-cell crusade made it contentious.

In 1972, Congress authorised $115 million for action and research towards the control of sickle-cell anaemia after President Nixon had spoken of 'sad and shameful neglect of this disease'. But in a Californian street I met a black engaged in a crusade against the sickle-cell crusade and the most conspicuous word on his collecting-box was 'genocide'. Racism and bad consciences about racism have charged with unhelpful emotions much of the discussion of this grievous but fairly rare clinical condition. For clear evolutionary reasons sickle-cell anaemia primarily afflicts black-skinned children.

Inside the blood cells, and colouring them red, is material called haemoglobin. It normally consists of many identical molecules. Molecular biology is all about such molecules, which are very small but in this case very intricate and precisely defined collections of atoms. The essence of life is nowadays seen to be the bustle of elaborate molecules of thousands of different kinds. Haemoglobin is just one of them; it is a working molecule, capable of absorbing and releasing oxygen as required. The atoms of haemoglobin are arranged in a variety of chemical units; these units in turn join together in precise sequences to make long chains; and finally the chains twist themselves into a sculptured shape. At first glance the shape is meaningless, but it carefully enfolds a trap for oxygen. This molecular 'lung' changes shape slightly as it takes oxygen in, or gives it up.

In the haemoglobin molecules of a child suffering from sickle-cell anaemia there is a chemical fault which makes them sticky. When they are out of oxygen, they can clump together; that is what deforms the blood cells. Treatment with a drug called cyanate, which reduces this tendency to clump, is showing promise in trials in New York. But the molecular fault is hereditary – which brings us to the genes.

A ploy out of Africa

When the father's sperm and the mother's egg unite, each contributes half the genes to the hereditary instruction-manual for a new human being. The genes are molecules, too, copied from generation to generation and fully repeated in most of the billions of little living cells which grow from the fertilised egg to make the human body. Although of immense antiquity and value, the genes are not like family heirlooms, to be locked away until the next generation; many of them play an essential part in the daily life of the cells. As the body grows by repeated divisions of cells, each cell switches on the genes appropriate to the part of the body where it finds itself, while irrelevant genes remain inactive. The genes consist of long chains of chemical units, but the kinds of units are different from those in haemoglobin. While haemoglobin is a *protein*, the stuff of the genes is *nucleic acid*, of a particular sort known as DNA. Proteins like haemoglobin adopt irregular shapes, but two matching strands of DNA are neatly twisted together in the famous 'double helix', discovered twenty years ago. Each cell has about two metres of

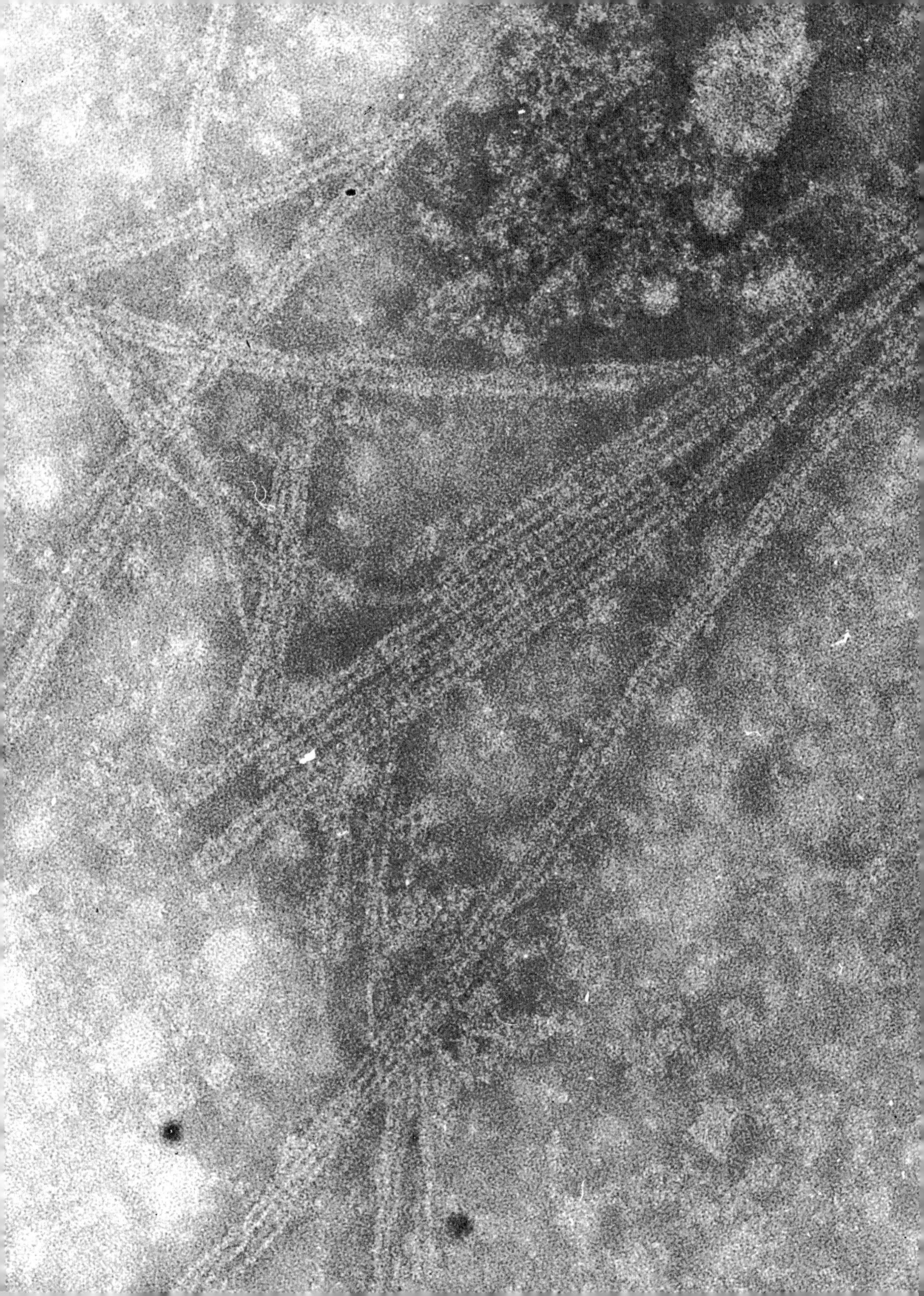

DNA packed into the minute nucleus, but not all of it is productive genetic material.

Heredity boils down to the fact that each productive gene instructs the machinery of the living cells to make a particular protein. The proteins so made are the executive agents of life, working molecules, each of which promotes a particular chemical process in the body. The message of the gene that specifies a particular working molecule is written in a simple chemical code. In a DNA chain four different 'letters', four kinds of chemical units (bases) normally labelled as A, T, G and C, compose a long, unbroken sequence of 'words'. Each word is of three letters and it names one of the twenty kinds of chemical units (amino acids) that go into the making of a working molecule.

If there is any 'secret of life' it lies here. The machinery of the cell translates the sequence of words in the gene into the sequence of chemical units in the working molecule. For example, a fragment of a gene reading 'CCT, AGA, CTT' translates into a portion of a working molecule with the following chemical units: 'glycine, serine, glutamine'. But I am not writing a chemistry textbook and the reader need not worry about what these names mean. All that matters is that, for any task in life, the working molecules have to be prescribed in this careful fashion by the genetic code.

The copies of the gene, needed in each cell of the body and in the sperms and eggs for the next generation, have also to be made with care. But the copying process is not absolutely perfect; indeed, if it were, life would not have evolved at all. Misprints occur in the genes and the name *mutation* is used both for a change in a gene and for its consequences. Many mutations are harmful and a mutation in the gene coding for haemoglobin is responsible for sickle-cell anaemia. A single, apparently trifling error of one letter of the genetic code (a T instead of an A) changes the meaning of the word, with the result that one incorrect chemical unit, the sixth from the beginning, goes into the corresponding chains of haemoglobin. That is what makes the molecule sticky.

Serious genetic errors like this, which are passed on to children, normally die out within a few generations, in part because people carrying the 'bad' genes are less likely to have children than those who do not. But the sickle-cell gene remains remarkably common among blacks – in some districts of Africa up to forty per cent of the population carry it. The reason for the sickle-cell gene's survival is that it can have a beneficial effect. Sickle-cell haemoglobin makes life uncomfortable for the organism that causes malaria during its residence in blood, and gives sickle-cell carriers added protection against the disease. When the sickle-cell gene occurs in a human population there is a record of severe malaria at some time in its past.

Each of us inherits two sets of genetic messages about making haemoglobin, one from mother and one from father. A sickle-cell carrier has the normal gene from one parent and the sickle-cell gene from the other; the carrier passes on one gene or the other to each of his children. If two carriers have a child, there is a one in four chance that he will inherit the sickle-cell gene from both parents; if so he will be a victim of sickle-cell anaemia. But half of his brothers and sisters will be carriers. That means they make both normal haemoglobin, which keeps their blood functioning, and sickle-cell haemoglobin which opposes malaria. In a malarial district of Africa, their better resistance to the disease offsets the death of the unfortunate baby with the sickle-cell disease. The 'bad' gene survives: most commonly where the malaria has been most severe. For the descendants of Africans now living in the USA, the absence of malaria means that no countervailing benefit can be

A blood sample taken from a three-year-old Washington girl is to be tested for the possible presence of sickle-cell haemoglobin, the defective version of the molecule made by the sickle-cell gene.

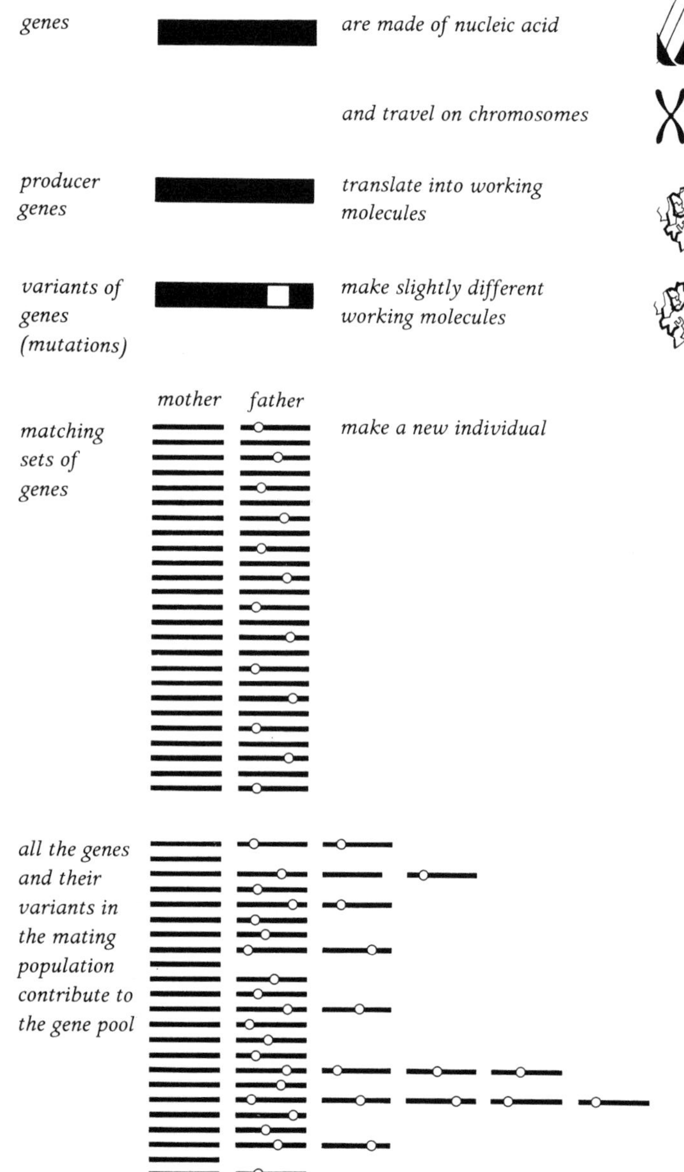

evolution is mainly a matter of variants of genes becoming commoner or rarer in the population

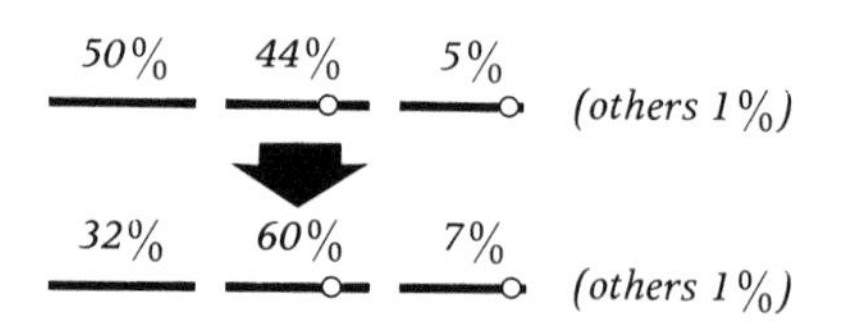

set against the deaths from sickle-cell anaemia. For this reason, and because of mating with whites and Indians, the gene is now much less common among American blacks than it was among their West African ancestors, and is still declining. That is evolution in progress today, among humans.

Darwin would recognise sickle-cell anaemia as a plain case of evolution by natural selection, which simply means that individuals vary and some are more likely to leave surviving offspring than others. Environmental circumstances determine which individuals these are. And because children tend to resemble their parents on account of the genes they inherit, a population little by little adapts itself appropriately to its circumstances – in this case either to malaria in Africa or to an absence of malaria in the USA.

The molecular alternatives

The most important discovery for modern evolutionary science, after the elucidation of the genetic material (DNA), came 13 years later, in 1966. With hindsight we can say that geneticists were slow to grasp just how great are the differences between individuals, whether human beings or other animals. It was something that Darwin stressed, descanting on the variability of humble barnacles. But geneticists had to find out the laws of inheritance from conspicuous characters controlled by genes, such as the visible configuration of a pea in a pod, smooth or wrinkled, or some manifest genetic defect like haemophilia or sickle-cell anaemia in humans. Misadventures in the pioneering days of blood transfusions disclosed genetic differences in blood types. The human body's 'immune response' for fighting infection is extremely sensitive to slight contrasts between its own molecules and invading molecules; it therefore attacked blood of a type foreign to itself. Since then, the blood group (A,B,AB,O and other variants) is routinely determined by testing a person's blood with other blood serum sensitised to react with particular types. Many years elapsed after the discovery of the blood groups, where there was plain variability in one of the few genetic systems that were accessible to ready investigation, before anyone seriously tackled the question of whether similar variability was to be found in other genes with no outward effects. Indeed some mid-century theories of genetics and evolution assumed that, disastrous mutations apart, comparatively few genes were variable.

Invisible differences between individuals took on a new importance with transplant surgery. Frantic efforts were made to find kidneys and other tissues genetically similar enough to the recipient's to give him a reasonable chance of not rejecting the transplant as if it were an infection. Tissue-typing laboratories are now part of the surgical scene, but some of the more heroic forms of transplant surgery have been curtailed precisely because in the end the genes win out. Heart transplants, for example, are ultimately rejected despite massive doses of drugs to suppress the immune response. The complete distinctiveness of any human individual from any other, except in identical twins, may be the bane of transplant surgery but it is the very essence of a species like man which has evolved in nature and retains unlimited possibilities for future evolution.

Simultaneously and unknown to one another geneticists in Britain and the United States decided to find out just how much variability really did exist among normal genes. They had to hand a crucial piece of knowledge that their predecessors were lacking, namely that the working molecules of the body are direct translations of the wording of the genes. They

Molecular alternatives revealed in the electric racetrack. In the technique of gel electrophoresis, materials are dragged through the jelly by a strong electric field, and molecules performing particular chemical work in the body are picked out by a special staining technique. If alternative forms of the molecules carry different electric charges, they reach different positions in the racetracks. The differences imply differences in the genes which prescribe the manufacture of the working molecules. They are remarkable tokens of the extent of genetic variability among individuals. The photograph shows little variation (in this case) between individuals of a population of flies, but variability among sea campions – the plant species shown at the right.

also had a new and convenient technique for sorting these working molecules, called gel electrophoresis. The idea is very simple.

A strip of jelly acts as an electric racetrack. Materials entering the jelly at one end are dragged through it by a high electrical voltage applied between the two ends of the strip. The speed at which each molecule travels – measured in millimetres per hour – depends on how cumbersome it is, how difficult is its passage through the jelly, and also on how much electric charge it carries. Some mutations, but by no means all, alter the electric charge on the molecule. For instance, an uncharged chemical unit may replace a charged one; this happens in the simple mutation that causes sickle-cell anaemia. When the haemoglobin from a carrier, a person with both normal and sickle-cell genes, is run through the electric racetrack, the normal haemoglobin will travel faster than the sickle-cell haemoglobin. After some hours it will be a few millimetres ahead in the race. The two forms of haemoglobin make two distinct bands across the strip and reveal themselves when the jelly is stained – the stain attaching to the racing molecules more readily than to the jelly itself. The same sort of separation occurs for contrasting forms of other working molecules.

In 1966 Harry Harris of University College London, working with materials from human blood, and Richard Lewontin and Jack Hubby of the University of Chicago, studying the materials in Drosophila flies, announced the same discovery. Alternative forms of working molecules – and hence of genes – were much commoner than anyone had suspected. Out of ten enzymes of human blood chosen at random for investigation Harris found three that had more than one genetic form; by 1972 the score achieved by Harris and other researchers who followed him had reached 20 out of 71. For their part, Lewontin and Hubby announced that, in five natural populations of flies, seven out of 18 randomly chosen working molecules were commonly variable. The researchers knew perfectly well that the electric racetracks were detecting only those variants among the molecules which possessed a different electric charge, so that their results greatly underestimated the actual occurrence of molecular variations.

Even so, its extent was astonishing, and all subsequent research of this kind has amply confirmed it. The working molecules in question have technical names like 'malic dehydrogenase' and 'alkaline phosphatase-7'; what they are and how they function need not detain us. It is enough to say that each of them plays a vital part in the intricate chemistry of life and that the nature of an organism is determined by the co-ordinated action of large numbers of molecules like these. Again, the technical name for the variability discovered in them is 'protein polymorphism'. For convenience, I shall refer instead to the 'molecular alternatives', meaning the abundance of alternative versions of genes, which are not just transient mutations but are scattered through a substantial part of the population.

Richard Lewontin, one of the discoverers, is an outstanding figure in current evolutionary research. Here is how he sums up the meaning of the molecular alternatives:

If all the individuals were the same, with respect to their genes, then evolution couldn't take place, there'd be nothing to change. For a long time, we didn't know how to measure or characterise these differences in the genes from one individual to another in a population. But now these new techniques like gel electrophoresis have made it possible to show just what the genetical differences are. . . . This is the way in which molecular biology and its new discoveries have made it possible to solve old evolutionary problems.

New Guinea people from two sides of a genetic dividing line. Those from Chimbu (upper photograph) seem to be descended from hunters. They halted the advance of farmers invading New Guinea from the east, whose descendants (below) live in nearby Goroka. Although the invaders were apparently lighter-skinned than the hunters, present-day skin colours are very variable.

An embarrassment of genetic riches – that is what the molecular alternatives represent. It is one of the consequences of molecular biology that have provoked, as we shall see, a crisis in evolutionary theory, going to the very heart of what biologists think they know about how evolution works and how we come to be the way we are. But important, less theoretical conclusions follow immediately from the discovery. Above all there are, in these electric racetracks, unprecedented intimations of individuality. Given such a range of choices, possibly among all genes, the number of different individuals that can be composed by shuffling the genes in sexual reproduction is to all intents and purposes unlimited – far greater, say, than the number of atoms in the universe.

Differences in only a few genes can be enough to distinguish one species from another. Yet a brother and sister will have many thousands of unlike genes. In that sense, a brother and sister – as they may have suspected – differ genetically more than the human species differs from chimpanzees. No two individuals, identical twins excepted, can begin to be even remotely alike, genetically. Above all, no individual, class or race can claim any monopoly of the 'best' genes or protection from the 'poorer' genes, any more than there can be such a thing as an 'average' or 'typical' person; whoever has the most-favoured versions of some of the molecular alternatives is certain to have less common versions of others.

History by heredity

Asking for a blood sample from a cannibal must take some nerve – even if he has reputedly reformed. Only a few legends tell what has been going on since the inhabitants of the island of New Guinea became isolated from the rest of mankind, before the Pharaohs ruled in Egypt. Geneticists from the Australian National University in Canberra, in alliance with anthropologists, archaeologists and linguists, are using blood samples to help disentangle this complicated but unwritten history. An extraordinary patchwork of humanity has spread itself across the mountains and forests of New Guinea. History and geography have divided it into more than a thousand distinct peoples, each with their own language, engaging in simple farming and often varying their menu with human meat.

Heredity has left its mark in genetic likenesses and differences between groups. Geneticists test them by determining blood groups and by closer scrutiny for other alternative forms of particular genes. Similarly, variations in language allow the linguist to deduce kinship between apparently distinct groups. Often the connections between languages which at first hearing are quite different come to light only by patient fieldwork of linguists who detect here a similar use of pronouns, there the same word being used for the same object, and so on. From such linguistic research evidence emerges of newcomers sweeping across the island more than once in the distant past and then being arrested and scattered like breakers on a sea wall. Neither linguistics nor genetics is reliable or self-sufficient; people can learn new languages and strangers can marry into tribes. But confidence grows when the linguist and geneticist arrive at very similar conclusions. Add the evidence of archaeologists, about the start of farming and other events in various parts of the island, and history begins to emerge from the misty past.

In the Eastern Highlands of New Guinea, the buried remains of ancient plants show that the first humans in the area were hunters living in an oak forest. The linguistic and genetic evidence reveals a complicated history for the hunters. An original population which arrived on the island perhaps twenty thousand years ago is now represented by small groups of descendants dispersed to different parts of the island. Two or three waves of other hunting peoples entered the Highlands from the west, the last wave having travelled along the north side of the island before swinging south into the Highlands. Then, in about 3000 BC, quite different people, lighter-skinned Austronesians, stormed in from the Pacific side. They were farmers who could cultivate bananas, sugar and root crops, and they kept pigs. They established a large beach-head on the east of the island, and pushed into the Highlands.

The invaders were finally halted by the hunting tribes at the Chimbu river. The linguistic and genetic distinctions persist to this day. Blood group B, for example, is twice as common on one side of a line on the map, among the former hunters, as on the other side of the line, among the invading farmers. The hunters learned farming from their enemies. The next big upheaval in life in the New Guinea Highlands occurred much later, about AD 1600, when the sweet potato was introduced. So well suited was it to mountain agriculture that it set off a population explosion among the highlanders. As the growing tribes dispersed into new territory, further micro-evolution occurred which is detectable in the genes.

When the investigators leave the broad, fairly clear-cut history for the fine details of kinship and distinctions among hundreds of different clans, more intricate genetic analysis is needed. For inquiring into seven related clans living near together in the Western Highlands of New Guinea, a group of geneticists took blood samples from everyone over fifteen. The samples were flown to Australia for testing. The usual ABO blood groups showed no significant differences. But other blood-group tests were more revealing, as were variations in the frequencies with which alternative forms of some working molecules showed up in the different clans. From this information the Australians were able to deduce genetic relationships between the clans, which well matched their verbal traditions.

Our changeable skins

The evolutionary history of Europeans is tricky too. At Stanford University in California an Italian geneticist, Luigi Cavalli-Sforza, has been collaborating with an archaeologist, Albert Ammerman, to find out about the spread of farming in Europe. Cavalli-Sforza, who is one of the world's leading authorities on human genetics, wants to know why Europeans turned pale and he thinks that human evolution in Europe is closely bound up with farming. Mapping the conclusions of different archaeologists, about the earliest appearance of farming in each part of Europe, has revealed a striking pattern. The agriculture invented in the Middle East 9000 years ago moved slowly across Europe, at a remarkably constant rate of one kilometre a year.

A new idea can travel by word of mouth, with the inhabitants of particular localities hearing about the invention from the occasional visitor from foreign parts; in prehistoric times, 'foreign parts' could mean just the other side of the hill. The other way to spread an important invention is for people who already have it to enter a new locality in substantial numbers, mingling with or even driving out the former inhabitants. Although prehistorians have favoured word of mouth as the means of transfer, as a geneticist Cavalli-Sforza doubts whether this was how farming spread through Europe.

The steady, wave-like progression suggests to him

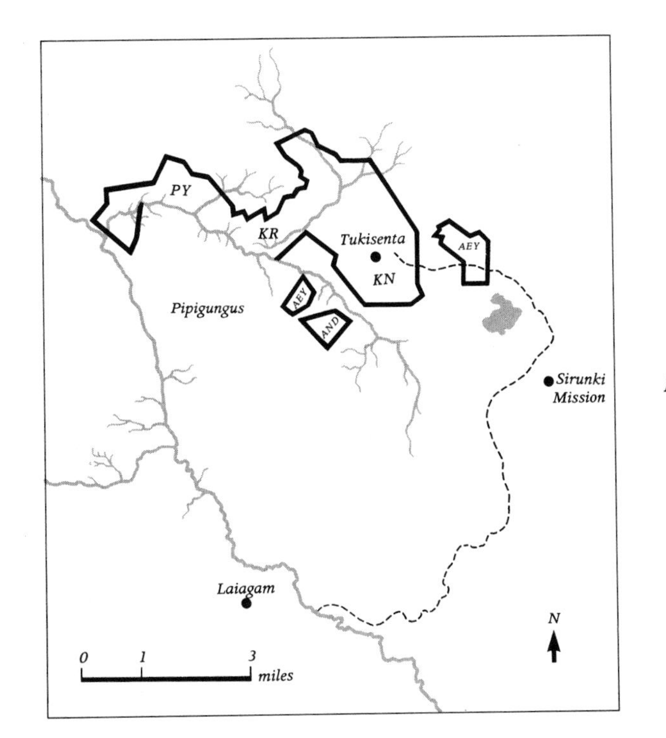

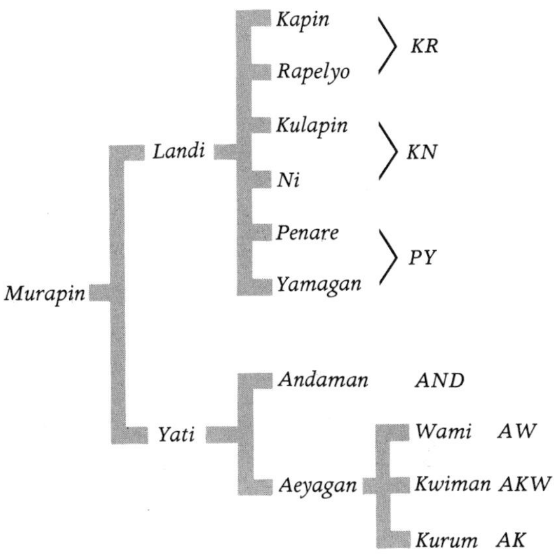

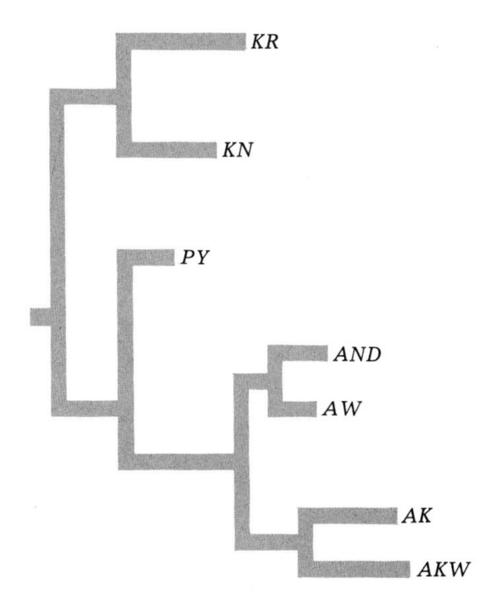

Teasing out genetic relationships between clans living only a few miles apart in the mountains of New Guinea. The map shows the territories occupied by clans of the Murapin group, designated by the abbreviations shown below. The upper lineage shows the relationships according to verbal accounts by the members of the clans; the right-hand lineage is that deduced by biologists at the Australian National University, from genetic similarities and differences. In this case, the lengths of the horizontal bars indicate the amount of genetic difference. (After P. Sinnett and others.)

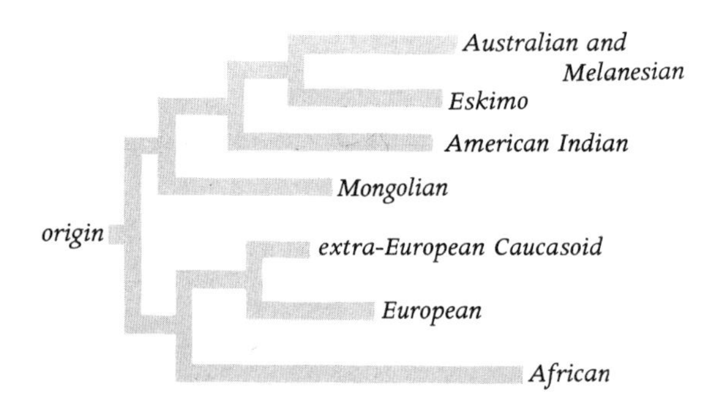

A human evolutionary tree (left) deduced by Luigi Cavalli-Sforza from detectable genetic differences. The 'extra-European Caucasoids' (Arabs, Jews, Indians) seem to be least different from the original members of our species.

a movement of people rather than of an idea. The original hunters of Europe would in any case tend to be outnumbered by the farmers whose populations grew on the basis of increased food supplies. If Middle Eastern farming did indeed travel in this fashion the evolutionary consequences would be a spread of the Middle Eastern genes as well. But those genes would become gradually diluted towards the north-west, as each generation interbred with some of the hunters whose lands they overran. In Cavalli-Sforza's view, the present-day people most like the former Europeans may be the isolated Basques of northern Spain. The genetic differences between the peoples of north-west Europe and the Middle East are slight, but as more markers of variations in human genes become available, Cavalli-Sforza hopes soon to have firm evidence for his story of the farmers' invasion.

As to why Europeans acquired white skins, Cavalli-Sforza wants more evidence from the archaeologists. He thinks it was an evolutionary accompaniment of the farmers' invasion. Farming peoples moving from the Middle East into wintry Europe were vulnerable to rickets – a weakness of the bones caused by lack of vitamin D. Hunting peoples, such as the Eskimos, get plenty of vitamin D from the animals they eat; but not the dark-skinned children of the early European farmers, living mainly on wheat and barley. Humans can make vitamin D in the skin but, in limited sunshine, skin pigment absorbs too much of the necessary ultra-violet rays. When farming was establishing itself in Europe, just a little vitamin D would give sufficient advantage, to any children whose skins were less brown than usual, to account for the evolution of the European white skin. So Cavalli-Sforza hopes for word from archaeologists that a few of the early farmers' children had rickety bones.

Mankind dispersed

An objective history of the evolution of the races of mankind, free from all false thinking induced by conscious or unconscious prejudices, is now offered by modern statistical comparisons of the genes represented in different human populations. Some anthropologists and sociologists think it reprehensible for the biologists to acknowledge that racial differences exist. But biologists probably do a greater service to the cause of racial harmony by teasing out the precise differences between human groups, and showing how slight they are compared with the differences between individuals within any one of the groups.

A human population, whether living in Stockholm or Swaziland, consists of a group of people living closely enough together to have a high chance of breeding together, and little chance of breeding with other populations, say in Nicaragua or Japan. Isolation in the past allowed different colours of skin, hair and eyes, and some facial features, to evolve in populations in different continents and regions. Dark skins, for example, are more resistant to damage from the ultra-violet rays from the Sun; light-skinned people are particularly liable to skin cancer in the tropics. But most of evolution in man, as in other animals, is not a clear-cut change, involving genes peculiar to particular populations. It is primarily a change in proportions, a shift in the relative frequencies with which different versions of the same genes occur in each population. Objective genetic analysis depends therefore on statistics.

In 1963, working primarily with blood-group statistics and a computer, Luigi Cavalli-Sforza and Anthony Edwards calculated the evolutionary tree for modern man based on the least amount of genetic change that was needed to produce the different patterns detectable in human populations. Refine-

It is easy enough to select faces that conform to old-fashioned notions of racial 'types'. Here are a Russian, a Formosan, a New Guinea tribesman, and a Negro. But to make such a selection is very misleading, as the following pages show.

The photographs opposite are all of Lapps in northern Europe. They show the great variability between individuals of one population. Below, for comparison, are people of quite different origins: Indian, Tibetan, American Indian and an Asian hill-tribesman.

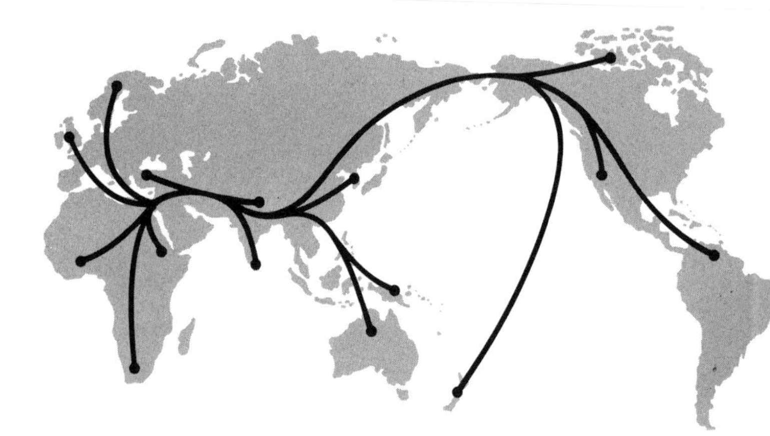

Genetic tests are much more reliable than appearances as indicators of closeness or distance between races. Here, a human evolutionary tree is projected on a world map to show possible routes of migration from an origin in the Middle East (after A. W. F. Edwards and L. L. Cavalli-Sforza).

ments in the past ten years have only tended to confirm their general account. The vast amount of new genetic information about humans, now coming from new molecular and immunological techniques, will no doubt clarify some details in the story. The conclusions fit well with ideas about human migrations deduced by archaeologists and anthropologists, and they dispose of a quasi-racist biological theory, still current in the 1960s, which said that our species was invented five times in different parts of the world.

Related benefits of Cavalli-Sforza's recent analyses are that they give a rough location for the single original population that gave rise to all the races of man, and a date when it existed. The 'Adams and Eves' of our species lived in or around the Near East 50,000 years ago. Present-day Arabs and Indians remain the closest, genetically, to those ancestors. By eastward and westward movements, two great divisions of mankind evolved: on the one hand the Asiatics, American Indians, Maoris and Australian aborigines, and on the other hand the Europeans and Africans. But there has been too little time for divergent evolution to proceed very far and the effect of modern travel has been virtually to end the isolation which had characterised human populations for many centuries. The differences achieved in the meantime between populations of black, yellow, brown and pink human beings remain literally skin deep. This conclusion of the geneticist finds strong support in the discovery of the immense genetic variability between all human individuals, represented by the molecular alternatives, far exceeding in extent any differences between the races.

What is the life game?

The marvels grow greater the closer we look at living things. And although modern biology has won some splendid insights and simplifications, it is being led ever further into a slough of complexity. Sydney Brenner of the Medical Research Council's Laboratory of Molecular Biology in Cambridge is engaged in the most elaborate effort in anatomy ever attempted. The intricacy of the living material that the Victorians dismissed as 'protoplasm' became evident only with the invention of the electron microscope. But snapshots of small sections of tissue under the electron microscope do scant justice to its organisation. To the electron microscope's ability to reveal details, Brenner adds the computer's capacity to digest and regurgitate on demand a vast amount of information.

This effort is dedicated to the nervous system of very small animals – nematode worms only a millimetre long, which normally live in the soil. The nematodes are cut into 20,000 exceedingly fine slices. One at a time, a long succession of electron micrographs are projected on a screen and an electronic pen is used to trace the animal's nerves in each picture. The pen automatically feeds information to the computer for storage. Brenner's reason for gathering and processing all this information is to compare the 'wiring' of the nervous system of normal nematodes with those of mutant nematodes that show peculiarities of behaviour.

About one hundred genes are involved in constructing the nervous system of a nematode and most of the mutations that occur affect the overall design of a section of the nervous system. This means that Brenner is dealing with genes of an organisational sort which regulate the wiring of the nervous system during the growth of the animal. After making major contributions to research on the molecular genetics of simpler organisms, Brenner has devoted seven years to preparing his new methods. That so much skill and effort has to go into the study of so simple an organism is a sign of how much biologists have still to find out about the exact organisation of living tissues. That even a nematode is so complicated shows how much evolutionary science has to explain.

If you were examining a jet aircraft, gleaming and packed with elaborate electronics, and someone suggested that the machine was there because the aluminium and iron and petroleum in the ground had spontaneously come together to assemble it, you would not be easily convinced. Yet the vastly more complex machinery of living things is the result of exactly such a process; only the chemical ingredients are different – carbon, nitrogen, water and so forth. Human beings, products of the Earth, make aircraft. In that sense the jet is just another by-product of life on our planet, like coral reefs and ant-hills. It is a minor achievement compared with a brain that demands to know how it comes to exist. The answer to the 'question of questions' is still in essence the same as Darwin's – that over long periods, through generation after generation, natural selection has favoured the inheritance of qualities well suited to the life of organisms.

Elsewhere in this book, we shall meet German molecular biologists playing with four-sided dice, and Japanese evolutionists using a computer as a roulette wheel. Games and the theory of games illuminate particular aspects of evolution. But there is a deeper sense in which life and evolution constitute a game. It is not a frivolous metaphor. The mutations of genes and the variability of individuals are products of chance, like a newly dealt hand of cards, but if that were all that happened there would be no life. The play of chance is modified by rules, just as in many card games, and the strategies that populations of living things unknowingly adopt, to make the most of the genes that chance has dealt them, are very like the activities of skilful players.

The prime rule which modifies the life game's abundant ingredients of sheer chance with an element resembling unconscious skill is natural selection. Some people talk of natural selection as if it were an agency capable of deliberate action, another name for God. Even some eminent biologists regard it as a special quality of living systems, almost a substitute for the 'life force' which was formerly thought to be necessary to breathe life into inanimate material. If it *were* confined to living things, life could never have begun. In fact, natural selection is no more mysterious than the rule in bridge that says a player must follow suit if he can – less mysterious in fact.

The rules of card games have to be thought out by intelligent beings, while natural selection happens naturally and inevitably. If you have living things – or non-living things, for that matter – that reproduce themselves reasonably faithfully but not too precisely, then natural selection occurs. As the resulting individuals are different, they cannot have exactly the same chances of reproducing themselves in further generations. But what is successful depends as much upon the environment as upon the innate qualities of the organisms. That is how organisms adapt to their environments, because the surviving genes at any stage tend to be suited to their circumstances. The messages from dead ancestors, saying in effect 'This works!', are what we live by. Most of our genes are almost unbelievably old, having been tested and proved in microbes, fishes and reptiles long before we began to borrow them.

A mixed 'strategy' of conserving what is valuable in hereditary information while always experimenting with new versions of genes has allowed organisms to stockpile, over countless generations, a great repertoire of 'tactics' for life. The variations make all the interesting individuals, races and species that we see around us. But the natural course of evolution, producing organisms more complex than any works of man, is only understandable given long periods of time. It is a slow process; if the Sun had blown up a thousand million years ago natural selection would have produced nothing more than microbes.

Evolution proceeds more plainly in other animals than in man. The next chapter deals with aspects of the life game in living species, and then Chapter 3 enters the domain of invisible changes in molecules, newly opened to evolutionary studies. After that we shall take a more historical approach, from the origin of life and the long but creative Age of Microbes (Chapter 4) down through the Age of Visible Animals to mammals and man (Chapter 5). The book closes with speculations about the future course of play.

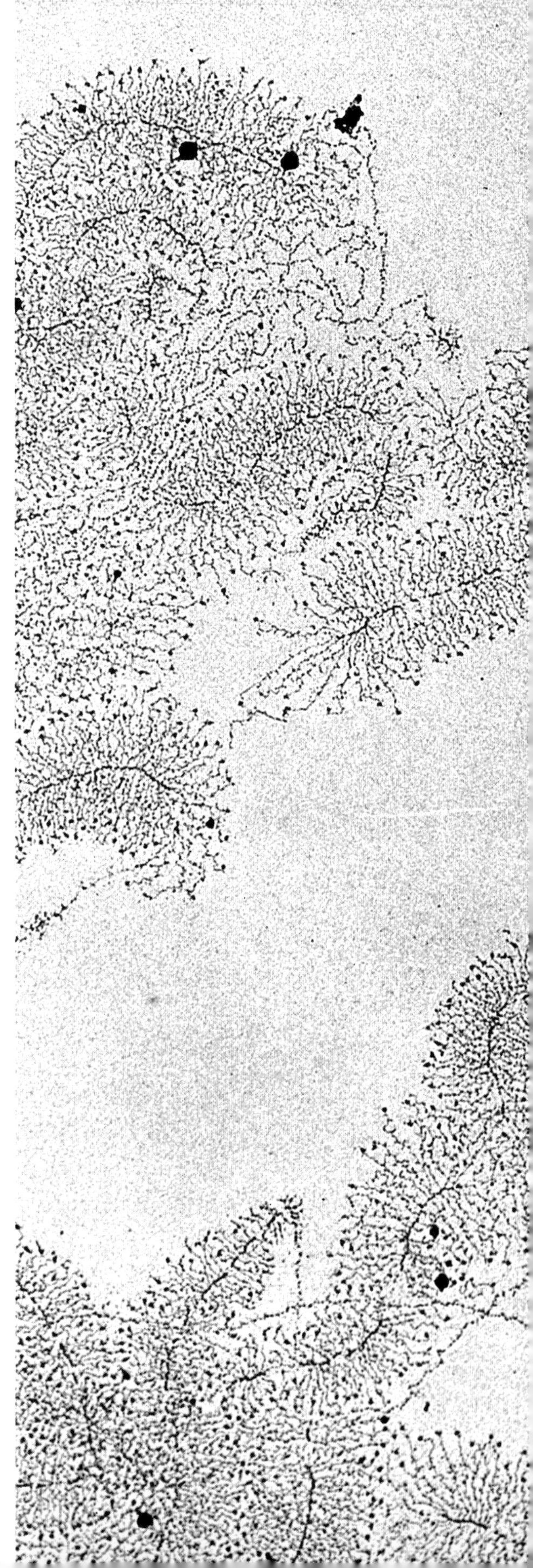

Genes in action. This remarkable photograph taken with an electron microscope shows genes 'sprouting' molecules that have transcribed the genetic messages and can thereby pass on the genetic instructions to the machinery of the cell. The axis of each gene is made conspicuous by working molecules that have attached themselves along it to help in the transcription.

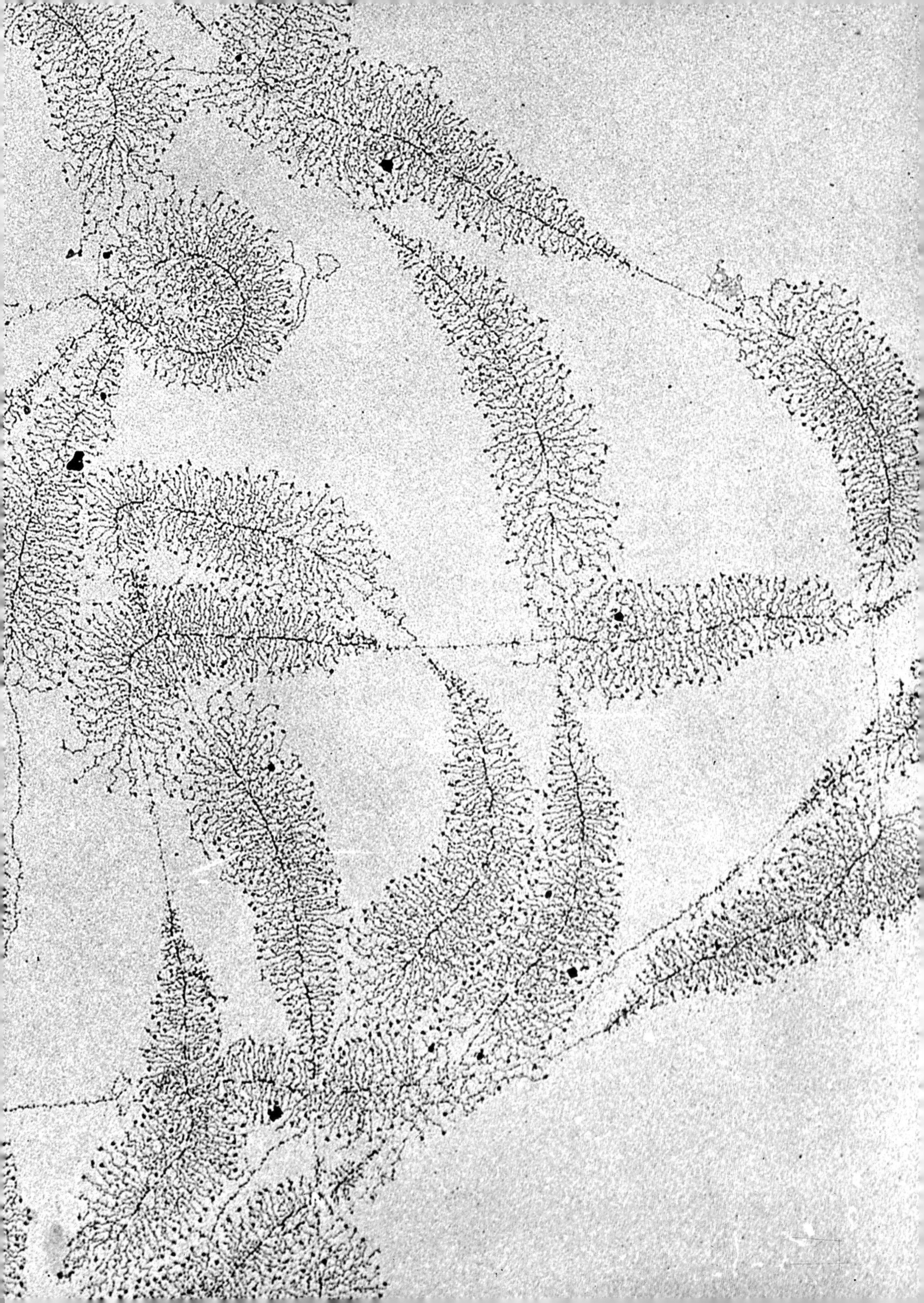

Chapter 2 How to Evolve

Flies are the heroines of the natural world's most remarkable case of high-speed evolution in the recent past. Drosophila, the so-called 'dew-loving' flies which feed and breed in fermenting plant material, have contrived in the Hawaiian islands to produce something like 700 different species in the course of a few million years. Many of the species have originated within the last 700,000 years since the largest island, Hawaii itself, arose from the ocean. In pockets of forest only a few kilometres apart populations of flies of the same species, identical under the microscope, show different proportions of alternative versions of working molecules, when these are run in the jelly of the electric racetracks. These shifting balances between different versions of the same gene tell of evolution continuing today.

Hampton Carson of the University of Hawaii communicated his enthusiasm as he expertly trapped flies with a glass tube, or swept the forest litter with his collecting net. After building a high reputation in St Louis as a biologist who combined theoretical probings with research in the field, he had settled in Hawaii so that he could devote his attention to the crazy evolution of the Hawaiian flies. We stalked flies among the ferns and bushes of the dripping forest, ankle-deep in mud and looking for insects so small as to be almost invisible at an arm's length. Carson was smearing the trunks and branches of trees with fermented banana baby-food. Without this bait some of the species were as hard to find as any elusive bird or mammal. In time, though, the tubes received their new occupants and the assortment of very different species illustrated the evolutionary diversity achieved by the flies in the great Hawaiian species-factory.

Although the 'big island' of Hawaii itself is smaller than Yorkshire, it is in the most isolated group of islands in the world. It also provides an extraordinary patchwork of living places for the flies. The tall volcanoes force the trade winds to release their moisture on the eastern side, and provide a gradation from warm rainy shores to cool misty uplands. Down-wind from the mountains lies semi-desert, in what geographers call the rain shadow, while on the far west of the island an eddy of the trade winds again produces damper conditions. On this climatic theme volcanoes have produced further variations. Every few years an eruption sends lava pouring down the slopes, engulfing wide tracts of forest and replacing them by barren expanses of chunky or meringue-like rock. These lava flows begin new stories of colonisation by pioneering species. In a great sea of frozen lava you will sometimes find a solitary plant newly rooted; it is not easy to imagine that after a couple of centuries the black rock will be hidden under a forest – a rich community of mutually supporting species.

The capricious flows of lava have another effect. They may by-pass pieces of the forest, or stop just short of them, thus allowing pockets of trees and animals to escape the holocaust. These surviving portions of forest are so much a feature of Hawaiian scenery that they have a special name – kipukas. They are islands within the island. In the kipukas isolated populations of plants and animals can evolve without regard to their relatives in other parts of the island. Thus the volcanoes continue the work of separating populations, which they have carried out on a grander scale along the chain of Hawaiian islands.

For forty million years a zone of volcanic activity has been punching up islands through the floor of

Hampton Carson in the wet forest in Hawaii, looking for flies. Among them, he is tracing an evolutionary explosion which has created hundreds of different species of flies, providing an exceptional opportunity for studying the origin of species.

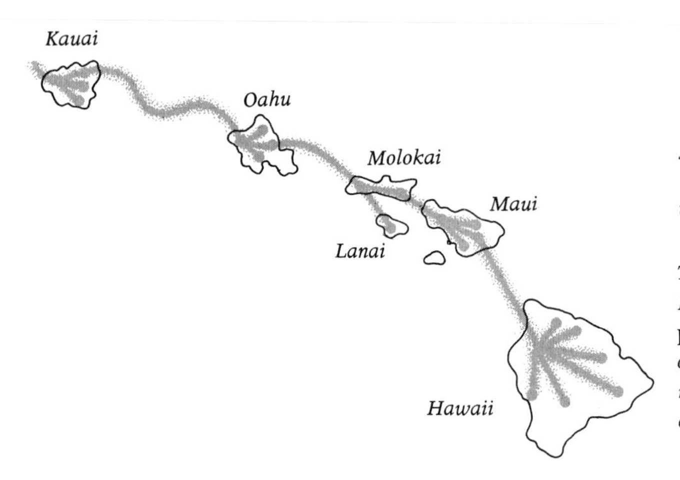

The Hawaiian chain of islands as a factory for new species. Kauai is the oldest surviving island and the others have erupted from the sea in turn. Plants and animals 'island-hopping' down the chain have become separated from their relatives, evolving into new species.

The great-grandmother of all the flies with spotty wings in the Hawaiian islands may have resembled this living fly, Drosophila primaeva, *from the oldest of the islands, Kauai. Patterns of bands on this fly's gene-carrying threads, the chromosomes, match the description for the ancestral pattern, deduced from comparisons of the chromosomes of the different species.*

the Pacific Ocean. The ocean floor is sliding north-westwards at a speed of a few centimetres a year, carrying each new-formed island away from the active zone. The islands then erode away – several islands have already disappeared except as sea-mounts on the hydrographers' charts. The present-day map of the chain is like a trail of diminishing puffs of smoke, streaming away from the volcano zone that now lies towards the south-east of the big island of Hawaii. From the point of view of flies that arrived by extraordinary chance from far-off lands to colonise the older islands of the chain, the picture has been one of new islands appearing successively to the south-east.

On each island in turn incoming flies have evolved into a variety of species; then, by rare events of wind and storm, occasional descendants have found their way to the next island to repeat the process. Sometimes species from younger islands are carried back to older ones, thus increasing the diversity there. Repeated separation has hastened evolution among the flies, as among plants and other animals. In Hawaii the humble tarweed has given rise to very varied species, some of them magnificent; similarly, birds now differing markedly in appearance have evolved from a single stock.

A lone representative of a species arriving to colonise new territory is a poor sample, in the statistical sense, of its ancestral population of widely varying individuals. From the outset, this genetic bias in the new populations encourages it to diverge from its ancestors, and make a new species. Another factor is the holiday from competition and natural enemies. The general shortage of inhabitants in the new territory gives the newcomers an opportunity to diversify, to fill some of the many vacant ways for life.

Phrases like the 'struggle for existence' and 'nature red in tooth and claw' have coloured many previous accounts of evolution. The evolutionary story does include harsh aspects, but they were much exaggerated. Diversity in the form of new species is more likely to appear when the living is easy. In Hawaii the absence of natural enemies also favours the evolutionary loss of protective mechanisms. There, the modest nettle has grown into a tree but it has also lost its sting.

Identikit for an ancestor

One of the Hawaiian species of Drosophila flies has a mallet-shaped head with its eyes on stalks; others, too, are bizarre in appearance or behaviour. Most of the species have been found and named only within the past few years, many of them by Hampton Carson's colleague at the University of Hawaii, Elmo Hardy. Often differences in outward appearance are slight, but a sharp-eyed entomologist can make small distinctions in the structure of wings, legs, sexual organs and so on. Some seemingly identical species are divided by differences of behaviour, especially in breeding behaviour. Sometimes an elaborate dance precedes mating; it is not done for fun but to prove to the female that the male knows the dance peculiar to her species, and is also sound in wind and limb.

The flies in one group of Hawaiian species have conspicuous spots on their wings. The males among them display their wings like little peacocks to impress the ladies during courtship. The wing-markings are another means of helping the female to see if the suitor is a member of her own species. Carson has made his closest study of these 'picture-winged' Drosophila of Hawaii. Their explosive evolution provides a special opportunity for the biologist to trace the appearance of new species – the principal expression of the

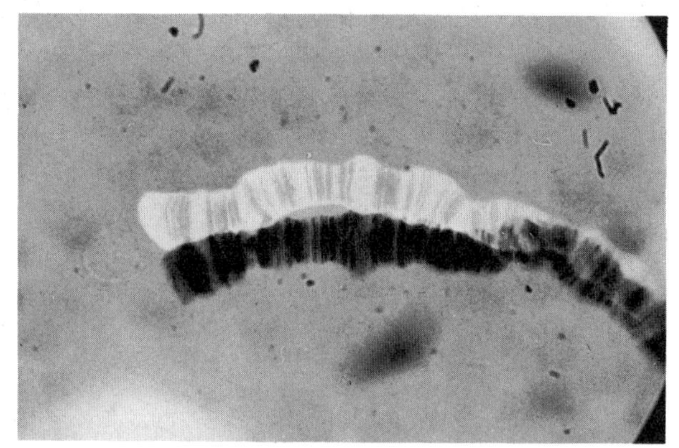

Chromosomes (left) from two different species of Hawaiian Drosophila compared under the microscope. The darker one is being viewed directly; the other is a photographic cut-out of the chromosome projected into the field of view. Notice the similar patterns of bands. Some of the many species of Hawaiian flies are just visible (right) at the tops of the pins in the museum cases.

process of life evolving on Earth.

Commoner species of Drosophila have been the favourite laboratory animals of the geneticists for many years. One reason is that the chromosomes (the gene-carrying bodies) in the salivary glands of these little flies are particularly large and easy to study. Carson exploits this characteristic, too, although for his purposes the chromosomal differences between species tell little about the actual mechanisms whereby new species come into being. Flies of very different species can have almost identical chromosomes, while similar flies may have different chromosomes. Nevertheless the chromosomes are very useful indicators of ancestral relationships between species. When he has removed the salivary glands from a fly and stained the chromosomes, Carson can make out patterns of bands through his microscope.

These bands blazon the genealogy of the flies. Recognisable sequences of bands occur in the chromosomes of the different species. More significantly, parts of the patterns are different – for example, reversed – from one group of species to another. Similar arrangements in different species imply kinship. An accumulation of changes becomes apparent in the chromosomes of the picture-winged flies. By his microscopic detective work Carson is able to trace ancestral patterns of chromosomal bands. Comparisons with the chromosomes of similar flies on the neighbouring island of Maui then reveal that the twenty-odd species of picture-winged flies in Hawaii are descended from nine different species in Maui.

The presumption is, therefore, that nine times in the past 700,000 years a pregnant fly has been caught in a storm, blown out to sea and across the channel to Hawaii, there to be dropped in a place where her offspring could survive. For every successful crossing there must have been hundreds of failures. The channel is twenty miles wide. But what of the ancestral flies that first appeared in the Hawaiian chain, after surviving an ocean crossing of some 2000 miles? Carson can give a chromosomal identikit description for the great-grandmother of all the flies with spotty wings in all the islands of Hawaii. A present-day species of fly living, appropriately, in the oldest of the surviving islands (Kauai) answers to that description and probably resembles most closely the female fly that founded the empire of new species in Hawaii.

Evolutionists suppose that a new species usually arises because a population of organisms, separated from its relatives, evolves to suit the particular environment in which it lives. It thereby becomes so different from its relatives as to be unable to breed with them if they are reunited. But in Hawaiian conditions, which are by no means typical of life on Earth, the process can evidently be reversed. Flies can separate into new species first, and only afterwards settle down to the serious evolutionary business of adapting themselves to their environment. Carson has a theory for this odd evolutionary behaviour.

Great population explosions may have occurred among the flies on the uncrowded islands of Hawaii. One result of a low death rate is that many peculiar individuals survive that would simply die in more normal circumstances. All sorts of strange mutants can thus appear among the flies. A day of reckoning comes when the flies outrun their food resources and the population crashes. But the scattered survivors would include some of the peculiar new forms; these could then establish themselves as new, distinct species by virtue of their genetic differences from other populations of flies. Carson has yet to convince his fellow biologists of the suggested reversal of the evolutionary conventions among his

Yellow eye-rings (left, below) are one of the marks of the Herring gull, imprinted on newly hatched birds and helping to determine their choice of mate. But the female Herring gull (upper photograph) has accepted a Lesser Blackbacked gull as a mate. She was hatched in the latter's colony during an experiment in which eggs were exchanged. Both birds are descendants of the same Russian gulls that spread in different directions around the northern world (right).

flies, and of his explanation for it. But bold thinking of this kind seems necessary to keep up with the undoubtedly break-neck rate of evolution among the flies of Hawaii.

The quest for a mate has its limits, even for the most persistent animal. The proposition that new species of animals appear when populations become sufficiently cut off from their relatives to be unable to interbreed with them is well borne out by the flies of Hawaii. A species is a compromise design suited to a variety of living places. A breakaway group will adapt more closely to its own locality and perhaps try out new ways of life. The test of whether or not it remains the same species is whether its members would still interbreed with their relatives if they were brought together again. An early priority in a new evolutionary step is a change in mating arrangements that gives genetic integrity to the new species. These 'isolating mechanisms', as they are called, vary wonderfully among the new species of Hawaii, from the dances mentioned earlier, to changes in the shape of the sexual organs or positions for copulation that would surprise even the Hindu folklorists. In other parts of the world, and possibly in Hawaii too, Drosophila flies use love-songs and scents as passwords for their species.

The changeling gulls

When dispersed animals follow a circular route that leads them back, after a long time, to reunion with their relatives, they may then decline to mate with them. More than a score of cases like this provide the clearest demonstration of geographical separation giving rise to new species. The evolutionists' prize instance is of gulls that spread both east and west around the world from a starting point in Siberia. The eastbound populations met the westbound gulls in Europe. The two groups turn out to be different species, although outside the zone of confrontation no one can put his finger on the map and say, 'Here is where the species diverged'. But there they are, on the shores of Europe: two crowds of very similar gulls which do not breed together.

The Herring gull came through Canada while the Lesser Blackbacked gull, with a darker back and wings, spread to Europe across Russia. Other visible differences between them are that the Herring gull *(Larus argentatus)* has yellow around its eyes and at the 'gape', the corners of the beak, while the Lesser Blackbacked gull *(Larus fuscus)* has ruddier eye-rings and gape. Although they constitute the classical case of 'circular overlap', the gradations east and west from the Siberian starting point are not completely continuous; apparently the birds were split into isolated pockets during the Ice Ages. Even so, the close kinship of the Herring gull and the Lesser Blackbacked gull is perfectly clear. If they wanted to, they could breed successfully together. In fact, by careless mating, natural hybrids sometimes occur when the species are living close together.

On islands off the coast of Wales the Herring gulls nest on the cliff's edge; a little way inshore, among the bluebells, live the Lesser Blackbacked gulls. The Herring gulls have learned to patronise the fishing boats and fish docks, where scraps are to be had rather easily, while the Lesser Blackbacked gulls find their own food in the farmers' fields or at sea. For the winter the Lesser Blackbacked fly to Spain, Portugal or North Africa but the Herring gulls stay in Britain. Thus the birds, although very similar in general behaviour, follow different ways of life and in their colonies, especially when breeding, the two species avoid each other. But as there is no physical bar to breeding, biologists had the opportunity to find the precise 'isolating mechanism' that normally separates the species.

Mixing up the eggs is one way to go about it. Starting in the early 1960s Michael Harris, an ornithologist now working for the Natural Environment Research Council, has exchanged hundreds of eggs between colonies of the two species on the Welsh islands of Skomer and Skokholm. The eggs look alike and the birds involved are unaware that they are hatching strangers. Harris marks the young birds with rings on their legs after they hatch. He has made repeated visits to the islands to follow the fate of the changelings. They can be picked out by the rings but also because they remain in their foster colonies and their unusual colouring stands out in the flock. They have fared well and adopted the ways of their foster parents rather than their true parents. Most notably, young Herring gulls brought up among the Lesser Blackbacked successfully make the winter migration with the rest of the flock.

The changelings also mate with their adopted flock and produce viable offspring that are themselves fertile. The behaviour which produces this result is revealing. The males of neither species are particularly choosy about their mates and if it were up to them the Herring gulls and Lesser Blackbacked gulls would today still be breeding as a single species. But the females will normally respond to overtures from a male only if he belongs to what she takes to be her own species. Nothing in her genes tells the bird what the 'correct' appearance should be; instead a well-known and very powerful psychological mechanism comes into play.

A newly hatched bird is 'imprinted' with the first object it encounters and becomes irreversibly lovestruck; in the normal course of events the object will be its mother but in artificial situations a young bird will treat a cardboard box or a flashing light as if it were its mother. Changeling gulls are similarly imprinted with the birds of the colony in which they hatch. In gulls, eye-ring colouring is crucial. Experiments in which the eye-rings of other gulls were altered, by painting them in the 'wrong' colours, have demonstrated the importance of this species-signal in mating. In the Welsh experiments the female changelings will accept as mates only male birds that look like their foster parents. The fact that the changeling female's own eye-rings are the wrong colour does not deter the more promiscuous males. So all that keeps the Herring gulls and the Lesser Blackbacked gulls apart is the female's preference for birds resembling her parents. It is a slight enough foundation for a new species.

The formation of new species punctuates the course of evolution. It lets differences be sustained that would otherwise be smudged out by interbreeding with the ancestral population, and it allows the new species to begin evolving in novel ways. A million years from now, Herring gull and Lesser Blackbacked gull may be completely unalike. Arithmetically, new animal species appear somewhere on Earth at a rate of perhaps ten a year, but as the process in each case takes thousands of years to accomplish biologists can only hope to identify cases of newly-formed species, as with these gulls, or else to spot separated populations that seem to be on the brink of the big split.

Except in a species-factory like Hawaii, for any one existing species or group of species the budding-off of a new species is an extremely rare event. Most of evolution is concerned with more subtle changes like those that have occurred in human populations, which fit species and populations a little better to the ways of life to which they are committed. The chief task of evolution theory is to convince the biologist

Overleaf, marine iguanas of the Galapagos, descended from South American reptiles that made their way by chance to the islands.

and the layman alike that the plants and animals we see around us, each wonderfully adapted to its environment, have acquired their elaborate forms and behaviour, little by little, in a wholly natural way.

Enchanted islands

'Nothing could be less inviting than the first appearance,' Darwin wrote of his arrival at the Galapagos Islands in HMS *Beagle* in 1835, with the South American mainland five hundred sea-miles astern. Other visitors have complained about these islands at the end of the Earth, but none has matched Herman Melville's purple periods.

> Take five-and-twenty heaps of cinders dumped here and there in an outside city lot – imagine some of them magnified into mountains, and the vacant lot the sea; and you will have a fit idea of the general aspect of the Encantadas, or Enchanted Isles. A group rather of extinct volcanoes than of isles; looking much as the world at large might, after a penal conflagration.

The alternative Spanish name for the islands is 'enchanted', not 'enchanting'. Melville referred also to their 'dreary spell'. Yet this very harshness, so unexpected in islands lying on the Equator and often enough the despair of seamen looking for water, is a substantial reason for the Galapagos being the most important acres in the history of biology. Here evolution took a striking enough turn to shake the confidence of the young naturalist aboard the *Beagle* in the Judeo-Christian proposition that God created all plants and animals in one grand population of the planet. That was the first step towards *The Origin of Species*, a quarter of a century later. With the hindsight made possible by Darwin's own theory we can tell why the islands are biologically so impressive.

In the modern idiom, we would say that the Galapagos Islands were like a new planet to be colonised. The volcanoes began bursting out of the Pacific Ocean about ten million years ago and were of course barren at first, except for marine life. A few living things found their way there from South America, carried on winds and currents like the raft of the *Kon-Tiki* expedition. Most that survived, against all the odds, the journey across the ocean must have been defeated at its end by the arid, unwelcoming lava of the islands. But a few plants did take root on the shores and others fared better in the damp mists and occasional showers higher up the slopes of the volcanic cones.

Animals that found food and a footing on the Galapagos evolved away from the ancestral populations from which they were now separated. Slowly and unknowingly they adapted their forms and their habits to exploit the meagre resources of the land and the riches of the sea. The marine iguana, the descendant of a South American land lizard, has learned to swim and to graze on sea plants; his cousin, the land iguana, suitably yellow in colour and leading a more solitary life, has become a diminutive reptilian goat with a stomach that can resist the thorns of a cactus. Most species of the Galapagos occur nowhere else on Earth. But that is true for many species and many localities and would not, by itself, have undermined Darwin's faith in Creation. More to the point were the differences occurring between corresponding animals living on the various islands, many in sight of one another.

Among the land birds, from a single founding parent species, finches evolved with great versatility into no fewer than sixteen distinct species scattered around the islands and following a variety of different ways of life. Some eat plants, some eat insects, and one plays the woodpecker, using a cactus spine to delve for

Galapagos tortoises mating.

insects in holes in the bark of trees. The giant tortoises which gave the islands their name vary too, in the sizes and shapes of their shells.

When the Vice-Governor remarked that he could tell from which island any tortoise had been brought, Darwin pricked up his ears. He had been carelessly mixing up his specimens from different islands, never dreaming that the islands would have been 'differently tenanted'; he quickly mended his ways. He examined the mocking birds collected by himself and his shipmates, and found to his astonishment that all the birds from one island belonged to one species and all from another to a different species. But he had already hopelessly muddled most of his specimens of the finches that were to make the Galapagos and himself jointly famous. Who can blame him? They are small birds, the males being black and the females brown. When you glimpse them flitting among the thirsty trees of the Galapagos it is hard to acknowledge the impact that such modest birds had on the human mind and its religious beliefs.

But Darwin took his mixed-up finches back to England, where he handed them over to John Gould for examination and classification. Gould confirmed a 'perfect gradation in the size of the beaks in the different species'. Some had massive nut-cracker beaks; others were small and delicate like tweezers; and there were many intermediate forms. Darwin commented:

> Seeing this gradation and diversity of structure in one small intimately related group of birds, one might really fancy that from an original paucity of birds in this archipelago, one species had been taken and modified for different ends.

The 'fancy' nagged him. Two years after he visited the islands, Darwin opened a notebook on the transmutation of species. He always acknowledged the strange animals of the Galapagos as the chief origin of his belief in the fact of evolution. It took him longer to hit upon the mechanism of evolution, in natural selection.

The dire opponent

When we went to the Galapagos Islands to commune with the strange birds and beasts my head was full of sophisticated stuff about molecules and populations, out of the laboratories of Australia and Japan. But here, I told myself as I sat on the rocks of Fernandina amidst a reptilian Greek chorus of marine iguanas, here if anywhere one should find a way of reconciling with real life those abstractions about the very small and the very numerous.

The iguanas, picturesquely ugly, all faced the equatorial Sun to keep their bodies cool. The carnival of reptiles ended seventy million years ago, but these dark-skinned anachronisms were as indifferent to my mammalian presence as to the Galapagos hawk that was looking for his lunch. Their collections of cells and working molecules were wrapped in a skin like dark sacking; its seams resembled stitching. Inside one of the iguanas, nerve cells responded to a dilution of the glucose in the blood. The animal stirred itself to drag its tail to the surf and swim away to graze on the seaweed. For a lizard this is an odd way of life, evolved by force of circumstance on these Pacific islands. But simply trying to visualise genes in action inside these sluggish animals built no strong bridges in my mind.

Better clues that came took an unexpected form. The bones of a dead iguana lay back from the beach, clean and intact, a perfect skeleton. In coming ashore we had contended with an exceptionally high tide as well as the usual surf. When at last we reached the headland where the flightless cormorants made their nests we found that the Pacific Ocean had swept

them away – nests, young, eggs and all – leaving only rings of black lava stained pale by droppings. The bereaved adults looked like scarecrows as they stretched their bedraggled wings, atrophied by evolution, to dry them in the sun.

At several places were sealions, the most joyful of mammals, lying fly-blown, dead or dying of a virus disease, while healthy youngsters played in the water a few steps away. The Charles Darwin Research Station at Academy Bay runs a hatchery for endangered varieties of the giant tortoises which had been altogether too tasty for visiting seamen. There we met the only known survivor of the tortoises of the island of Pinta: an adult male. Unless visitors to his island stumble upon a mate for him, the Pinta tortoises will follow those of Santa Fe and Floreana into extinction.

On Isabella we visited a colony of blue-footed boobies. The boobies earn their disrespectful name because, lacking, like many Galapagos animals, any self-preservative fear of the human visitor, they are ludicrously easy to catch. But seeing these birds in action over the water makes you think again. A flight of boobies drops like a stick of bombs into the sea, and the cool curving world of certain fishes comes to a sudden terminus. Adults and young birds stood in the scrub of the colony surrounded by dead chicks and unhatched eggs. The boobies practise an austere form of population control; the mothers' eggs hatch at different times and when food is scarce the younger chicks cannot compete for it with their older siblings. They are allowed to starve. In that season the place was littered with white balls, fluffy like cotton.

Each of those small corpses was a masterpiece of nature, grown with sublime precision from a mixture of its parents' sub-microscopic genes – genes that would have instructed it how to grow feathers, how to breathe, how to fly, how to bomb the fishes of the Humboldt current. For these birds the game had ended almost as soon as it began, in unrequited cries for a share of the food. Their unique combinations of genes would not participate in the elegancies of avian courtship and make new generations.

Premature death will extinguish excellent systems of genes and working molecules unless their internal intricacies are matched to a favourable environment and sustained by a reasonable quota of good luck. Here is one factor that links the molecular models and statistical theories of the laboratories to the course of play in the wild. The genes and working molecules, the elaborate machinery of cells and organs, new tricks of animal design or behaviour – all are ploys in the game of life against death. The dire opponent acts through unfortunate individuals on the fate of genes in the population.

Rules and players

The essence of the life game is that organisms should produce offspring which are similar to, but not identical with themselves. That is the only cardinal rule of the game and the practical rule of natural selection is immediately implied by it. If organisms are not identical, they will not have equal successs in living and reproducing; consequently the more successful repro-producers will come to predominate. The result will be a gradual accumulation of hereditary messages advising the descendants how to live.

Everything else that can be said about evolution deals with the particular circumstances of life on Earth and strategies for the game that appeared by chance and were stockpiled by natural selection. Perhaps even death should be counted a special characteristic of life on Earth. It is not strictly necessary for

A large blue-footed booby chick and a younger chick lying dead. One of the volcanic landscapes of the Galapagos (following pages) suggests the harsh, semi-arid conditions with which life of these islands has had to contend.

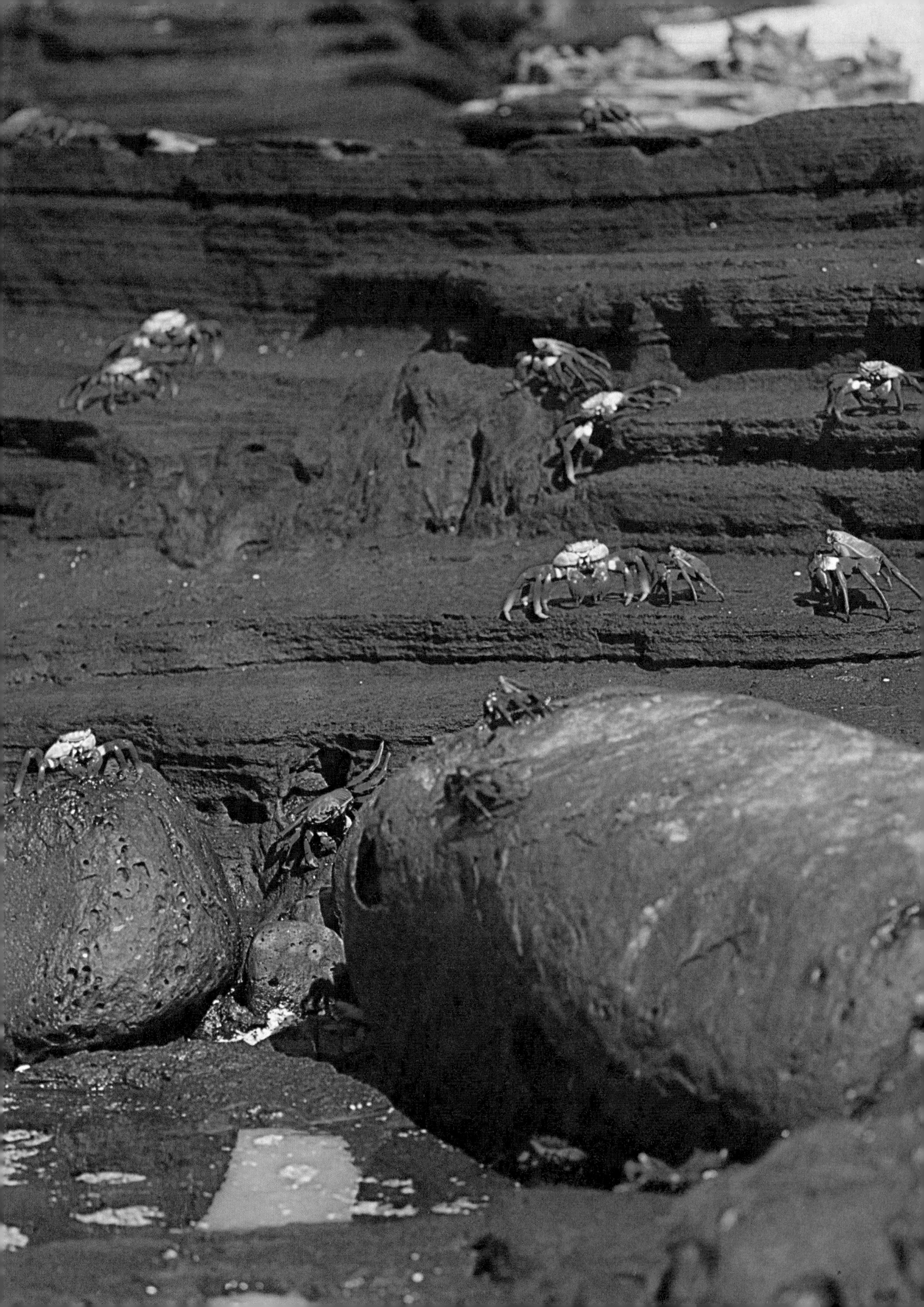

Galapagos crabs.

natural selection which only requires variations in success in reproduction. On an infinitely large planet this process could continue without any need for death. In practice the resources of our planet are limited and one organism has to die to make room for another. And it undoubtedly speeds up the game. The shorter the interval between generations, and the briefer the span of one species until its extinction and replacement by another species, the more quickly can natural selection bring about changes.

Life on Earth has found it opportune to organise itself in mortal packets which can die without ending all life. The existence of individuals and of species is so conspicuous a fact of life that we take it entirely for granted. One could imagine more fluid or more casually organised forms of life existing on the moons of Jupiter. But it is not easy to think of any forms better suited for evolutionary success; presumably if there were a better way, life on Earth would have adopted it, rather than what we see around us.

Everything goes in units. The gene is a unit; it maintains its identity through many generations but is prone to mutations. It can be combined in random fashion with other genes but you cannot have a soupçon of a gene – it is either there or not. Whether, in its particular gene-combination and environment, it is fully expressed is another matter. Genes are wrapped up in the packages we see as individuals; the individual is a unit that will succeed or fail as a whole. The population is a unit wrapped up in its sexuality; so is the species, the collection of populations that will interbreed with one another (given reasonable opportunity) in distinction from other species with which it will not interbreed. Ernst Mayr describes the merits of this arrangement:

Organising organic diversity into species creates a system that permits genetic diversification and the accumulation of favourable genes and gene combinations without the danger of destruction of the basic gene complex.

Which of these various units – genes, individuals and populations – can be regarded as the 'player' in the game against death? A strong candidate is the pool of genes embodied in a population of a particular species living in a particular area. Sexual breeding within the population maintains a certain cohesion in the gene pool so that, in principle, almost any gene could be combined with any other. The individuals produced by breeding are the various 'plays' or 'moves' of the gene pool, which may succeed or fail. A play succeeds if the individual in question produces offspring that themselves survive and breed. Any other outcome entails a loss of gene-combinations. The object of the game is to keep the gene pool in being.

At first sight the individual does not rank as a player nor is his death of any consequence on its own account. Indifference to the fate of individuals is indeed well shown by the insects and fishes that produce vast quantities of fertile eggs, very few of which survive. Premature death, or a failure to reproduce for other reasons, matters only to the extent that it may shift the balance between the different genes in the population. The individuals and indeed the whole population are factories for making copies of the genes. The implication might seem to be that the chicken is just the egg's way of making another egg. But matters are not quite as simple as that.

Genes are expendable too. The population carrying the gene pool must be sustained even if the price is the loss of particular genes and their substitution by others. Every gene could be renewed, in principle, and still leave the organisms recognisably the same. To have a place in the organism the genes must be useful to its way of life. And here the individual re-emerges as a vehicle for genes and the proof of their effectiveness in organising the growth and life of the whole organism. The 'chicken-or-egg' argument about the primacy of one aspect or another of life dissolves: there is a triumvirate of the gene, the individual and the population, any of which is competent only by virtue of the other two. Which of them counts as the 'player' depends upon what aspect of the life game is under consideration. Far from being a weakness in the metaphor, this ambiguity about who is the player straddles precisely the areas where biologists have made best use of the notion that life is a game.

Nature abhors perfection

The individual is in the firing line and progress in the game depends entirely on the fact that life is embodied in vast numbers of variable individuals which may either reproduce or fail to reproduce without the population as a whole being at much risk. Only in the individual is a set of genes expressed; the individual, too, in his successes and sufferings, his sex-life and his death, puts his genes at hazard and allows natural selection to do its worst. Yet individuals cannot evolve in any biological sense. Selection operates only on individuals; evolution occurs only in populations.

The idea of evolution occurring in populations rather than in individuals is crucial to the modern understanding of the process. Suppose that a promising new gene – a beneficial mutation – first appears in an individual called John. John is mortal and the new mutation can become important only if John passes the gene in question on to his descendants. Only half of his children and only one in eight of his great-grandchildren will have the gene and they are mortal, too. There is every likelihood, in fact, of the gene being lost along the way. Only when quite a number of individuals possess it is it a serious candidate for adoption in the population. If, over the course of many generations, the gene becomes commonplace in the population, then the population will have evolved in that respect; the fact that the mutation occurred first in John, a thousand years before, is quite irrelevant. In his day, he was only an individual with a mutant gene. Everyone has mutant genes, but the establishment of an important new gene is a comparatively rare event. Most of evolution most of the time is just juggling with the relative frequencies of long-existing genes and combinations of genes.

But why does natural selection never complete its work? Why does the population not evolve to become 'perfect', with all individuals splendid but identical, so that evolution ceases? With no options left open the population would be vulnerable to the first substantial change in its environment. Yet it might seem that this was the inevitable outcome of the game. For many years the belief was widespread that all individuals would be 'perfect' if they did not all carry some harmful mutations that natural selection has not yet eliminated. The fact is very different: nature abhors perfection. The enormous number of coexisting versions of hereditary material, in the recently discovered molecular alternatives, shows how far real species are from any such condition.

Several mechanisms, apart from new mutations, operate to maintain the genetic variability that is the hallmark of the evolutionary healthy species. One is the fact that the environment is never the same from year to year, or from place to place in the species' range. Other mechanisms have various technical names but all have to do with the different ways in which a particular gene performs in different associations with other genes. There is no clear verdict on whether it is 'good' or 'bad', so diversity persists.

The interplay of individuals and species raises more questions. Why do we go on bothering with sex? It is an extremely inconvenient way of reproducing yourself; you have first to find and win a mate and then waste enormous quantities of sperm and eggs to achieve the occasional fertilisation. Most species operate this way, and there are obvious benefits in the continual remixing of genes that sex affords. Moreover, the sheer awkwardness of sex gives natural selection another lever for more rapid adaptation of the species; an animal must be reasonably efficient to have any real chance of reproducing itself this way. For the species sex is unquestionably a good thing, but natural selection does not operate on species, except through its individual members.

Now look at sex from the individual's point of view. An excellent ploy would be to give it up. If you do away with all the rigmarole of courtship and mating, and generate offspring by virgin birth, it all becomes much simpler and more reliable. Any animal possessing this ability has a clear reproductive advantage over those that do not, so it ought to be strongly favoured in the short run by natural selection, whatever long-term disadvantages there may be. There are plenty of cases of simple animals giving up sex, and even some fishes have done so. But they are a small minority of species and the problem is to explain why most species have been faithful to sex. The answer must lie in the benefits to the species as a whole, but no one has yet managed to explain how the obvious reproductive advantage of virgin births to the individual is suppressed for the well-being of the species.

The theory of games

Other posers about individuals and species are less baffling. One is altruism. Why should an individual share his food with his companions, or risk his life for them in more desperate ways, thereby hazarding his genes? The benefits to the species are plain enough, if lives and genes are thereby saved and co-operative ways of life can develop. But at first sight natural selection would seem to work the other way: reproductive advantage would lie with the selfish individual and any genetic tendency towards altruistic behaviour should quickly be extinguished for that reason. Yet among mammals in particular, and in man most of all, altruism has evolved very strongly.

The explanation seems to be game-like. Suppose that an individual – let us call him John again – possesses a gene that makes him fairly altruistic. The object of the game is to preserve that gene. What strategy of behaviour evolved by the gene will best preserve it? Each of John's own children has half his own genes, so roughly half of them will have the altruistic gene. It is then worthwhile (from the gene's point of view) for John to gamble his own life at up to 50-50 odds to save one of his children. Roughly half of John's brothers and sisters will also have the altruistic gene, so again altruistic behaviour up to a fifty-per-cent risk of death is self-justifying genetically. Only one in four of John's nephews and nieces, and only one in eight of his first cousins, will have the altruistic gene, so, in theory, his altruistic zeal should be correspondingly moderated. Strictly speaking he should let his sister-in-law drown unless she is pregnant. But in practice the risks of saving life will seldom be overwhelming and the advantage will time and again outweigh the personal jeopardy it invites. The altruistic gene will prosper.

Family life and social organisation, in man and other animals, have been described as games in which the individual is trying to win the best possible prospects for his own genes. Co-operation, which is not the same thing as altruism, is favoured by evolution in any game where it gives the genes of each participant a better chance of survival. But the strategies for males and females, for parents and children, for relatives, neighbours and colleagues all differ. The result is the intricate balance of rivalries and coalitions that we experience every day, albeit profoundly modified by the learning capacity and social sensitivity of the human brain.

Evolutionists now do more than offer loose descriptions of games. They are making increasing use of a strict branch of mathematics known as the theory of games. It originated in the 1940s, not as an aid to winning at poker but for dealing with sterner contests of a game-like character such as commercial competition, diplomacy and military strategy. The idea of trying to apply it to life occurred to a number of mathematicians and biologists, but the first serious attempt at it was made by Richard Lewontin, whom we met earlier as one of the discoverers of the molecular alternatives. Lewontin turned to the theory of games in the hope of extending the mathematical theory of evolution. Up till then it could predict changes only within an essentially stable population exposed to a more or less constant environment. For the instabilities in populations, which create and extinguish species in the more dramatic events of evolution, only vague, non-mathematical theories were available.

Lewontin described the game played by a population of plants or animals against 'nature' and he posed the question of which strategy the population should adopt when confronted with capricious 'moves' by nature – that is to say, changes in the environment. To answer such a question in detail would depend, of course, on how the progress of the game was to be scored to show when the population became better off, or worse off. But quite general conclusions can be drawn about the merits of different kinds of strategies. The two most promising strategies are those known to games theorists as 'maximax' and 'maximin'.

The maximax strategy is the one that will give the best result of all, but only if nature 'plays' a certain way. It is the hardened gambler's strategy of win-all or lose-all. Maximin is much more cautious; it is the strategy that gives the most favourable – or least unfavourable – results in the worst moves that nature can make against the species. In practice a species may adopt a mixed strategy in which some of its constituent populations are playing maximax and others are playing maximin. Mathematically there is an ideal blend of strategies. In their several environments, the populations may be emphasising very different gene combinations and this may create a situation in which a new species can appear.

But the population is not really a conscious player, deciding what is best for the survival of its members. The game proceeds only because groups appear within the population, the gene composition and breeding activity of which corresponds to one strategy or another, whether maximax, maximin or stupid. In the course of time the groups unwittingly adopting appropriate strategies will 'win' in the practical sense that they will survive while other groups die out. The species that become extinct will include 'gamblers' whose maximax strategy fails when environmental circumstances change.

Thou shalt not kill

When two animals of the same species fight they almost always avoid doing any serious injury to each other. This is so even though the animals have

Heraldic lions serve to symbolise three kinds of actions for an animal involved in a fight: fighting 'conventionally' (with snarls, cuffs and so on), escalating the fight and retreating.

teeth or horns or claws capable of doing great damage. There are exceptions – frequent in particular species such as the Siamese fighting fish and by rare mishap in normal species. These exceptions are far outweighed by the restrained fighting behaviour of the great majority of animals. For the evolutionist, this has been one of the conspicuous features of life very difficult to explain by natural selection. The most illuminating use of the theory of games in biology so far has been to solve this puzzle.

Contests between animals of the same species are often about serious issues bearing on the survival of the individual and of his genes – notably in fights over food or a mate. Yet the animal's behaviour, programmed in his genes, generally keeps fighting down to the level of conventionalised snarls, gestures and token bites; if tempers should rise, the fight is quickly broken off by one or other party. Why are animals so restrained?

Obviously it is better for the species as a whole if its members do not go around killing one another. But again the problem bobs up: natural selection works on individuals rather than the species. A very fierce dog which assailed his companions ruthlessly and grabbed the choicest pieces of meat and the choicest females for mates would seem well placed to leave more surviving offspring than the other dogs. The genes that favoured this ruthless behaviour would then become commoner in the population. But plainly this does not happen as a rule.

John Maynard Smith is a former aircraft engineer who switched careers and is now an evolutionary theorist of high repute working at the University of Sussex. The way he uses the theory of games to attack the problem of restrained fighting is not entirely straightforward, because it involves an important new evolutionary concept – of animals in a population

conventional fighting

escalation

retreat

Fights and scores compared, as a way of determining the 'evolutionarily stable' behaviour for animals in conflict with their own kind. The various strategies are described in the text. Each fight is shown in two phases; what an animal does in the second phase may depend on what his opponent does in the first phase (adapted from J. Maynard Smith).

Scores: *certain victory* +2 *loser, uninjured* 0
half-chance of victory +1 *risk of serious injury* −10

Golden pheasants fighting.

	vs Dove	vs Hawk	vs Bully	vs Retaliator	vs Prober
Dove scores	+1	0	0	+1	0
Hawk scores	+2	−10	+2	−10	+2
Bully scores	+2	0	+1	0	+2
Retaliator scores	+1	−10	+2	+1	−10
Prober scores	+2	0	0	−10	−10

Fighters and scores. The figures are the scores assigned to an animal adopting the strategy named at the left, when fighting with a fellow-animal adopting the strategy shown at the top. The strategies are explained in the text.

showing 'evolutionarily stable' behaviour. By that Maynard Smith means the behaviour which, if it is practised by the majority of the animals, cannot be improved upon to the individual advantage of a rogue animal behaving differently. Restrained fighting is an evolutionarily stable form of behaviour, which accounts for its commonness in real life. To demonstrate this, Maynard Smith compares the strategies open to an animal in a fight.

From three possible actions by the animal – conventional fighting (snarling, etc.), escalated or dangerous fighting, and retreat – five plausible strategies can be compounded, to which Maynard Smith gives nicknames:

Dove: fight conventionally; retreat if the opponent becomes alarming.

Hawk: escalate the fight at once and go on regardless.

Bully: escalate the fight at once but retreat if the opponent retaliates.

Retaliator: fight conventionally but escalate if the opponent escalates.

Prober: fight conventionally but escalate if the opponent also fights conventionally.

Maynard Smith next follows the practice of games theory and assigns scores to the outcome of the fifteen kinds of fights that could occur between two animals, when each can choose one of the five strategies. An animal that is certain to win the fight earns a score of +2; when his strategy has a fifty per cent chance of winning his score is +1, and when the animal certainly loses but escapes uninjured his score is zero. But if escalated fighting continues to the point where the animals run a serious risk of suffering injury, both of them are given a big negative score.

The resulting table can then be examined to see which strategies are evolutionarily stable. *Bully* never gets hurt, but in a population of bullies a *Hawk* would do better, so *Bully* is not evolutionarily stable. Only one strategy – *Retaliator* – fits the bill. In a population where this is the general policy no animal adopting one of the other four strategies will do any better, in the long run, in the competition for food or mates. For example, a rogue *Hawk* in a population of *Retaliators* will meet retaliation in his first fight and will have a fifty per cent chance of suffering serious injuries – not a promising future for his genes. Meanwhile the rest of the population, among themselves, are confining themselves to conventional fighting and their chances of injury are negligible.

Fumigating a micro-island off the Florida coast. The scaffolding supported a tent for the purpose of destroying all the small animals on it. This drastic ecological experiment was to study how, afterwards, newly arriving species came into balance.

Most animals have in practice adopted the *Retaliator* strategy. Thus the theory of games shows, in this case at least, how genes benefiting the population rather than the individual can be sustained. The behaviour of the individual in relation to the rest of the group is what is under test by natural selection.

Nothing evolves alone

A plant attacked by caterpillars may evolve the manufacture of repellent chemicals or poisons that will deter the insects. In turn, the caterpillars will evolve means either of avoiding the plants in question or of resisting their chemical warfare. In other cases, plants and insects have evolved together in more co-operative ways. A flower advertises its presence to attract insects and feeds them with nectar, in order that the insects should carry the pollen to other flowers. The invention of flowers (quite late in the history of life on Earth) and evolutionary changes in insects that made the most of the plants' generosity provide one of the most striking examples of 'co-evolution'.

It is always easier to think about one thing at a time and evolution theory has concentrated on following the fortunes of single species. The life game is made much more complicated – and more fascinating – by the interactions between species. These are the very stuff of evolution. Nothing lives alone, as ecologists never cease reminding us. Other species are conspicuous elements in the environment of each species, both as sources of food and as potential agents of death. Success or failure of the genes depends on what species are relying on the same limited resources and also on subtle relationships between predator and prey, the eater and the eaten.

An animal must conserve its food resources or it will perish, but ecological arithmetic can play as big a part as genetic ingenuity. A predator that kills rather too many of its prey will find the prey harder to catch so that the predator will tend to starve and the prey to recover. Nor is it entirely true that animals never kill wantonly and never irresponsibly injure the vegetation in the natural habitats. Elephants and their relatives have been damaging vegetation for millions of years and may in fact have contributed to the thinning of the African forests which created conditions in which man evolved.

Natural woodland normally contains more than one species of large trees. In a free-for-all you might expect one species to displace all others because it is particularly well suited to the conditions of soil and climate. But if a species of tree begins to dominate, its parasites and pests will have such a field-day that many of the dominant trees will be killed, leaving room for others to spring up; thereafter the damaged trees recover because their pests now have a less easy life. Thus interactions between species encourage diversity.

When two species follow exactly the same way of life, competing for precisely the same resources in the same place, one or other will usually be extinguished. The 'exactly' and 'precisely' are important though. In plenty of cases species neatly dovetail their requirements. Two closely related species of Drosophila flies in Hawaii breed in the sap oozing from the injured bark of a particular shrub. But one species lays its eggs on the shrub itself; the other in the puddle on the ground where the sap has dripped from the tree.

But how many species can inhabit a given area? Important lessons both for evolutionists and for ecologists come from discoveries about the number of different species of similar animals that an isolated territory of a particular size can support. In

Two islands (or enclaves on land) can support, between them, a greater number of different species than can the larger island formed by uniting them.

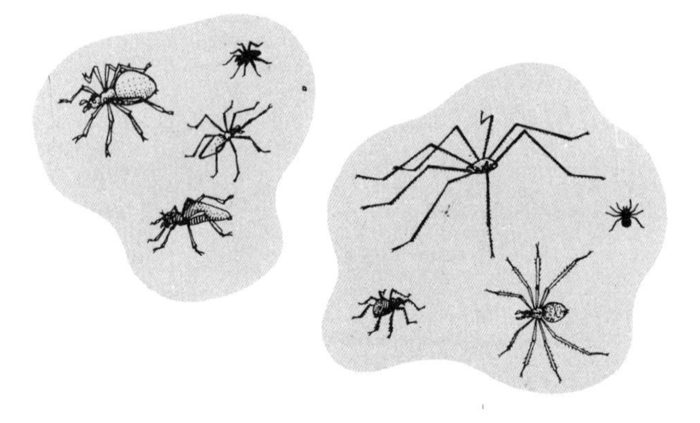

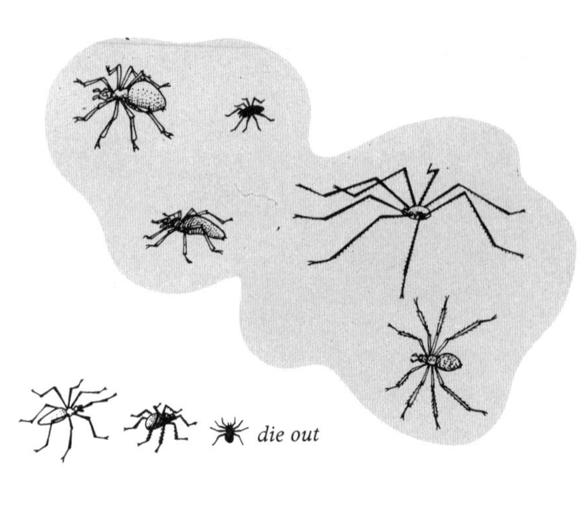

1966-7 two biologists, Daniel Simberloff of Florida State University and Edward Wilson of Harvard University, began an unusual ecological experiment in the Florida Keys. They took over six small islands which were little more than clumps of mangroves in shallow water. They covered these micro-islands with plastic tents and fumigated them with methyl bromide, thus killing every insect and every other animal on them. Within a few weeks fresh animals were arriving and after two years most of the islands had recovered fairly stable populations, and each was approaching the same number of species as before the extermination. In detail, the results bore out predictions about the way species would at first succeed or fail in colonising an island, irrespective of one another, and then begin to interact, reducing somewhat the total number of species.

Closely related to this process of species coming into balance is a more general rule. It deals with the number of species than can co-exist in islands or other isolated habitats of different sizes. Roughly speaking, a tenfold increase in area is needed to double the number of species. If isolated territory of one square kilometre can support ten species of a particular group of animals, ten square kilometres would allow only about twenty species the opportunity for life. This rule is not necessarily very exact or reliable, but it helps to explain why small areas, like the Hawaiian islands, are disproportionately rich in species.

Similar groups in similar environments, animals feeding on plants, animals feeding on animals, and parasites and pests, tend to be similarly organised. How far these similarities persist, going back into the distant past when very different kinds of animals were dominant, is one of the current challenges to research. So far evolutionists have only just begun to come properly to grips with the fact – appalling in the complexities it implies – that nothing evolves alone and that what happens to any one species can be fully understood only in relation to the adventures of the other species around it.

To the individual species the appearance in its neighbourhood of an enemy not under its control looks like a chance event, as does a volcanic eruption or a change in the climate. Chance also figures in orthodox evolutionary theory in creating the reservoir of variations in the gene pool and in the individuals, and in special events such as the arrival of small numbers of animals on a new island. But all this chance is subordinate to the ever-guiding hand of natural selection. That is the main message of the New Darwinism, which I have tried to convey in this chapter. So massive is the evidence for it that it must be broadly correct. But is it the whole story?

Mixed vegetation on English downland photographed by infra-red light. Most plant life on land has been modified by human interference, and what is regarded as the 'natural' cover very rarely is so.

The golden mother. The nearer of this pair of Midas cichlids of Nicaragua, seen with their prodigal output of newly swimming fry, has turned to gold. She represents the ability of living things to try out different characteristics. In this case, natural selection probably balances a certain dominance of golden individuals over other members of their species, against their greater vulnerability to predators. But are all evolutionary changes in organisms so nicely judged?

Chapter 3 The Molecular Heresy

Brian Hartley calls some small experimental tanks in Cambridge 'our Galapagos islands'. In them he and colleagues at the Medical Research Council's Laboratory of Molecular Biology bring about the evolution of simple organisms. Bacteria consist of only one living cell apiece and they grow and reproduce themselves very rapidly. In a few days they go through as many generations as large animals do in a thousand years. A small culture also contains vast numbers of individuals. These characteristics make bacteria particularly useful for evolutionary experiments.

The Cambridge experimenters grow bacteria of a kind known as *Klebsiella aerogenes* on quite unsuitable food. They supply them with a synthetic alcohol that scarcely exists in nature and this diet makes life exceptionally challenging for the bacteria. At the start of each run, they can grow only very slowly, but after a week or so a dramatic growth-rate increase occurs. The bacteria have found a new strategy for life. Nor is this the only event; later, other surges in the population occur, each marking another step in the bacteria's evolution. Sometimes a dose of ultra-violet rays, which encourage genetic mutations, helps the evolutionary process along.

To preserve 'fossils' of this history of bacterial evolution, the experimenters take samples of the bacteria from the tank at each stage, and store them. With the analytical resources of modern biology, Hartley and his colleagues are able to follow the course of play in great chemical detail and see how the bacteria have solved the chemical problem posed to them. The steps in their evolution are all cases of natural selection favouring unusual forms among the bacteria which are better able to cope with the unusual food and which therefore multiply more rapidly.

Time and again, evolution follows an almost identical course. As a first step the bacteria increase fivefold their production of a particular working molecule – a key enzyme called RDH that enables them to digest the synthetic alcohol. Extra copies of the gene that instructs the manufacture of RDH then appear in some bacteria. This bolder step enables the bacteria further to increase production of RDH. Before long, after succeeding steps, the bacteria are making quantities of RDH that would be ridiculous in normal circumstances – so much of it, in fact, that one-fifth of all the protein in their cells consists of the enzyme or about twenty times as much as in the ancestral bacteria that first arrived in the 'Galapagos island' of the laboratory tank.

Anyone who finds the repeatability of this sequence of evolutionary events too much like predestination, or perhaps resents the experimenters playing God to the bacteria, may be reassured by one unexpected ploy of the little organisms. In the normal way, each bacterium lives as an individual and is sooner or later swept out of its tank by the flow of the nutrient liquid itself. The evolutionary innovation that took the experimenters by surprise was the appearance of bacteria that were sticky. They produced clumps that resisted eviction from the 'island' and even succeeded in clogging the experimental machinery.

Interesting and amusing though all this bacterial evolution was, Hartley and his group were after something more. The enzyme RDH enables the bacteria to survive, but the fact that it is needed in great quantities before the bacteria will actually prosper shows that this way of dealing with the artificial food is inherently poor. The Cambridge molecular biologists wanted a mutant to appear with a novel working molecule that was more efficient than RDH. It is worth stepping back to see the reasoning that inspired their experiments.

Unbridled genes

In 1960 Brian Hartley considered the effects of certain poisons on the action of various enzymes from various organisms and tissues. The poisons in question were akin to military nerve gases, which kill in a terrible way by blocking crucial working molecules involved in the control of muscles by nerves. It turned out that a group of different enzymes serving different functions were all affected by the poisons, which implied a certain chemical kinship. The full power of molecular biology was brought to bear on some of the enzymes in question, both to analyse the chemical units making up their protein chain and to see, by X-ray techniques, how the chains were folded in making molecules that worked.

Only then did the similarities between the enzymes become fully apparent. Among the enzymes that help mammals (including ourselves) to digest protein are trypsin, chymotrypsin and elastase. They have different jobs to do but they all have similar molecular structures. They also resemble a protein-digesting enzyme found in bacteria. Just as zoologists deduce a common ancestry for all animals that have backbones, so molecular biologists see an evolutionary relationship in the common features of these enzymes. Presumably one molecule has evolved into others serving different purposes.

But now imagine the gradual, unit-by-unit transformation of one enzyme into another. There would be an intermediate stage in which the molecule would perform neither its old job nor its new job properly. Sickle-cell anaemia is a harsh illustration of malfunction in a working molecule due to a chemical change in its composition. So how can new enzymes with new functions ever evolve? By having extra copies of the gene in question – that is the evolutionary trick which Hartley emphasises. In the course

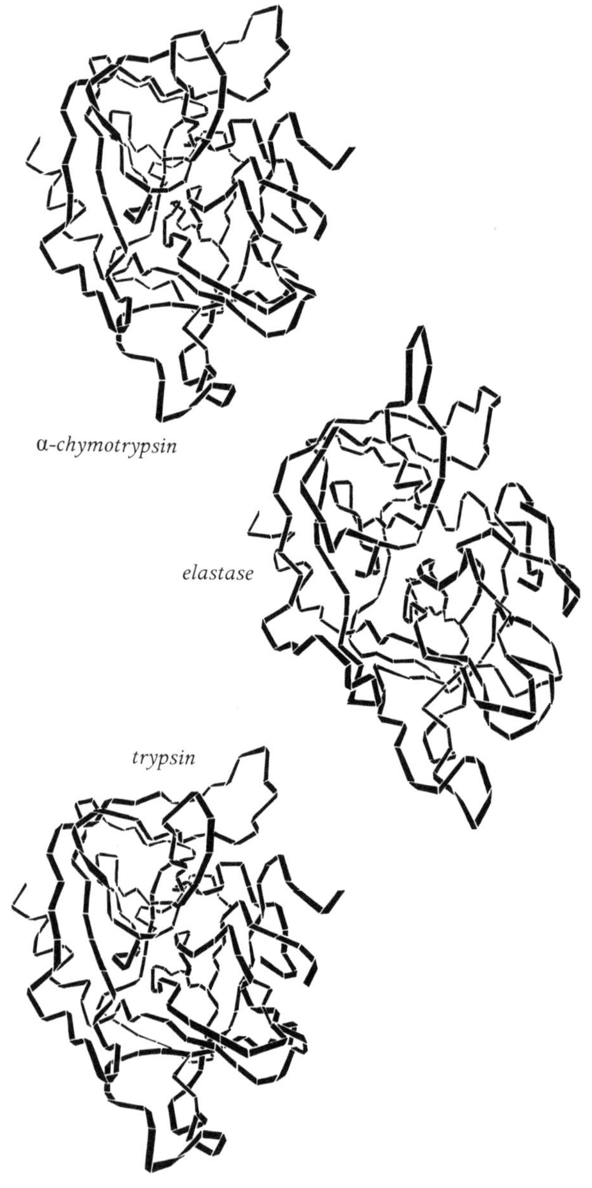

How gene doubling can permit new molecules to evolve.

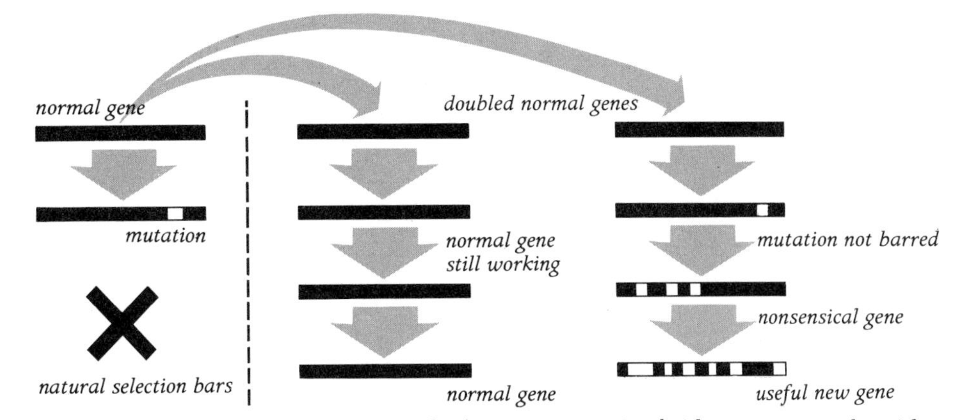

of evolution in his bacterial experiments, the organisms do indeed come up with extra copies of the gene responsible for making the essential enzyme RDH. The initial benefit is simply to increase the supply of RDH. But spare genes also represent great potential for evolution. If one copy of the gene can supply the organism's needs for the essential molecule – looking after the housekeeping, as it were – a second copy is in theory free to go through a long succession of chance mutations without any dire consequences for the organism. Thus the second copy is relieved of the very strict policing by natural selection to which its 'brother' is exposed.

Nearly all the molecules made by such unbridled mutations of the extra gene would be nonsensical. But, given time, a chance accumulation of mutations could very well produce a molecule that worked, but was different enough from the ancestral molecule to play a new part in the life of the organism. The working molecules employed by organisms today arose, perhaps, in this fashion. The chief aim of the evolutionary experiments with bacteria at Cambridge has been to encourage the bacteria to go through this process and thereby evolve a new molecule that will short-circuit its digestive process for the artificial food.

The experimenters had little reason to be optimistic. Although living things possess, between them, an enormous variety of enzymes and other working molecules, more than 3000 million years of life on Earth have gone in accumulating this biochemical repertoire. The odds against producing a novel enzyme to order in the laboratory seemed so long that a leading American molecular biologist has been building a huge automated production line for bacteria. In it ten thousand simultaneous cultures can be forced to evolve by peculiar nutrients or drugs.

Hartley's group at Cambridge persevered with their much simpler trials for four years, time and again achieving only increased amounts of the old, inefficient enzyme. The protracted gamble paid off early in 1973. Some of the bacteria adopted a new evolutionary course and came up with not one but several novel enzyme molecules, closely related one to another, which digested the synthetic alcohol more efficiently.

These experimental studies so far support the idea that the duplication or multiplication of genes is an important tool of evolution. Its major significance is that it allows nature to experiment with the structure of an existing gene without the strict constraint of producing active working molecules. In this way 'silent evolution' may be going on continuously in organisms that possess multiple copies of genes.

Doing without fossils

Just how deeply the discoveries of molecular biology strike into the vitals of traditional biology may appear from the following game for scientists. It has never been played in quite the way described, although it could be. I call the game 'Doing without fossils'.

Find a molecular biologist who is a good protein chemist. Preferably he has never handled a fossil and knows no more about natural history than is necessary to catch some moths, and some frogs' tadpoles. In these days of specialised education, this criterion will not be hard to meet. Apart from his little bit of hunting, he needs to go shopping for some pork, rabbit, chicken, fish, a cauliflower and some baker's yeast. His partner is a radiochemist capable of measuring the ages of rocks by the radioactive processes that have occurred in them since they last melted. He is supplied with pieces of rock (with no fossils, of course) from the major mountain chains of the Earth. The players must have the tools of their trades – including a Beckman Protein Sequencer – and a few months to spare. They can then proceed to deduce the course of play in evolution in the distant past.

The molecular biologist sets to work, taking from the various specimens he has caught or bought one of the working molecules. Among the easiest to extract and to analyse is cytochrome c. It occurs in all plant and animal tissues as one of a set of working molecules, first invented by microbes, for dealing with oxygen. He has to find out the precise chemical composition of the cytochrome c from the different species, as a sequence of more than 100 chemical units (amino acids) in the long chain of the molecule.

When he has listed, in order, every one of the chemical units that make up the cytochrome c in each of the various samples, the molecular biologist can start comparing them. Most of the units match up precisely in every sequence, which is not surprising because the chemical task of the molecule is exactly the same in all species. But no two sequences are identical and the molecular biologist can write down the number of chemical units by which each of them differs from one selected cytochrome c – say, the pig's:

	(chemical differences)
Pig	0
Rabbit	4
Chicken	9
Frog	11
Tuna	17
Moth	27
Baker's yeast	46
Cauliflower	47

Meanwhile, the radiochemist has discovered the principal episodes of mountain-building in the last sixth of the Earth's history (older mountains have worn away and do not look like mountains any more). These periods are:

	(million years ago)
Asia, Africa, S America, Europe	700 to 550
N America, N W Europe, Greenland	460 to 400
Europe, N America	350 to 225
Andes, Rockies, Alps, Himalayas, etc.	100 to 0

If the radiochemist is puzzled about what all this has to do with the cauliflower that his partner has been so busy with, a charitable umpire may suggest to him that perhaps there is some connection between momentous geological events and the history of life on Earth. So he takes his numbers to compare them with those of the molecular biologist. They waste a lot of time trying to fit the unfittable, until they decide to leave the cauliflower and yeast out, as far as the mountains are concerned. Then, when the numbers are set out on suitably proportional scales, they look like this:

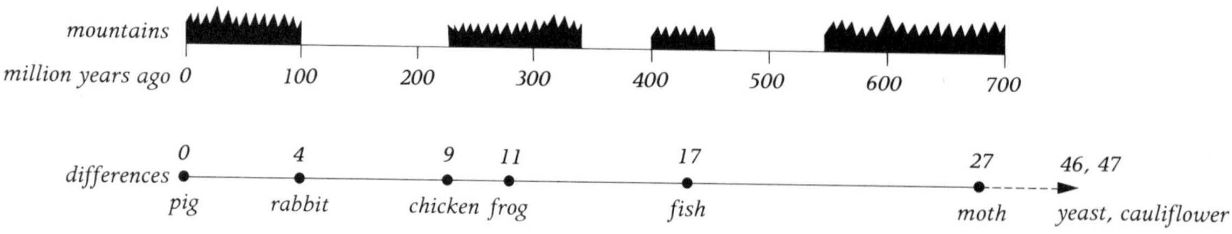

The fit is really quite good, but what does it mean? Certainly not that the tuna, for example, evolved 450 million years ago; we are dealing with modern species. Since everything is being compared with the pig, the pattern means that the *youngest common ancestor* of the fish and the pig lived about 450 million years ago. So our players, after all their labours, can draw a diagram of ancestors:

The radiochemist may then remark diffidently that biology is not his field, but he's noticed that frogs don't have wings and that yeast is a good deal crumblier than cauliflower. These observations prompt the molecular biologist to go back to his lists of chemical units. He then reports that the cytochrome c of chicken and frog differ by 11 chemical units, while cauliflower and yeast differ by 50 units. In other words these pairs of specimens are really as far removed from each other as they are from the pig. So the diagram is redrawn:

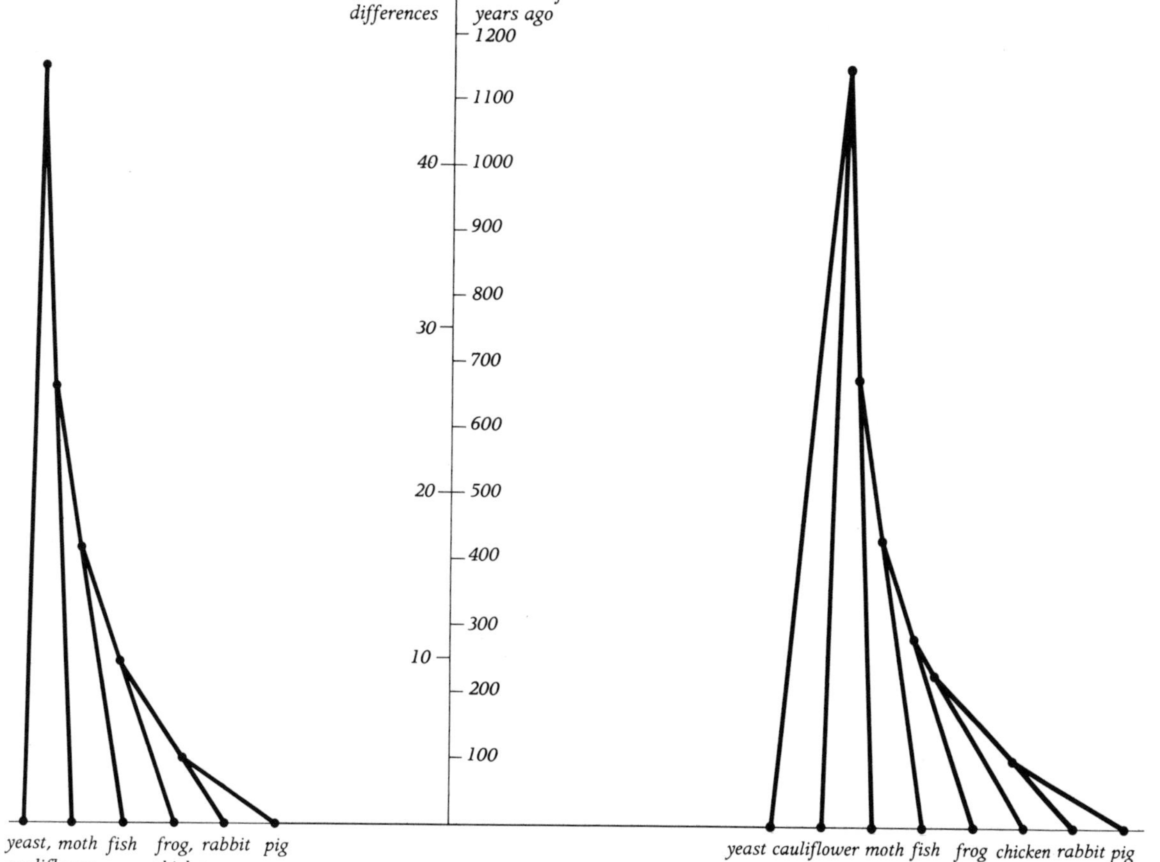

While the players stand back to admire their handiwork, the reader is invited to compare it with the diagram opposite. It is the evolutionary 'tree' as deduced by half a dozen generations of dedicated fossil-hunters, who have braved treacherous cliffs, mountain cold and desert heat to recover their specimens from the rocks in order to give us a picture of animal ancestries. Only the dates that are nowadays attached to geological periods, previously known only as 'Silurian', 'Devonian' and so on, depend on the sort of radioactivity tests as the players used in the game. Otherwise the approaches were completely different.

This game of 'Doing without fossils' is in several respects improperly contrived, in order that it should be as casual as possible. I did not trouble our molecular biologist with catching reptiles, for example, so the historically important reptile line is missing from his diagram. He could not, in any case, have bought dinosaur meat at the butcher's. More culpably, I chose the pig and rabbit as mammals with little difference in their cytochrome c, otherwise there is no fit to the most recent mountains. And, technically, I have skipped all sorts of tricky points about interpreting dates of rocks, about the divergence of animal stocks before new kinds of animals appear, and about how the counts of molecular differences are confused by more than one change occurring at one place in a molecule.

Molecular biologists have indeed produced much more elaborate evolutionary trees, simply on the basis of chemical comparisons of molecules (cytochrome c and others) between many living species. Their work (omitting the mountains but matching the dates) involves snakes, kangaroos, pumpkins, and many other common and rare living things. It does not matter to them that you cannot extract cytochrome c from a fossil. Indeed, one of the most striking merits

Part of the conventional evolutionary tree derived from painstaking examination of visible characteristics of fossils and living organisms. The comparisons of invisible molecules gave (on the previous page) essentially the same picture.

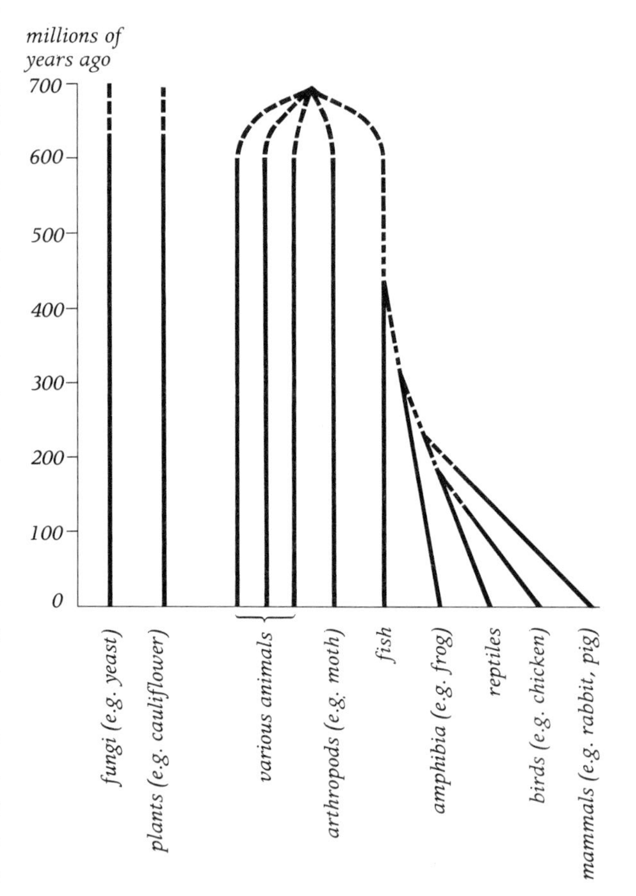

of the diagrams produced by molecular biologists, as in our game, is that they say something about the divergence of plants, fungi and animals, for which no fossil record exists.

Clockwork mutations

Even our rough-and-ready game of 'Doing without fossils' is workable only because of two momentous conclusions newly drawn about the course of evolution. The first is that there is some link between mountains and the major departures in evolution that create the branching points in the evolutionary tree. The connection is indirect and has to do with the movements of continents – but that fascinating aspect is postponed until Chapter 5. The second conclusion, which biologists find particularly disturbing, is that working molecules evolve, accumulating changes at a roughly constant rate. Molecular biologists could not make plausible evolutionary trees from their counts of molecular changes if cytochrome c had evolved in an erratic way – for example proceeding in the fishes faster than in the insects.

The molecular evolution with which we are dealing here is somewhat different from what was sought for, and discovered, in the Cambridge experiments with bacteria. That was the evolutionary invention of new enzymes – working molecules with new functions. In the case of cytochrome c, the molecules keep precisely the same function, whether in the pig or the cauliflower. But the changes in their chemical composition mean that mutations have gradually accumulated in the gene that instructs the manufacture of the molecule. The fact that all the surviving versions work to the satisfaction of the species relying upon them is by itself a strong hint that many mutations can occur which have no great practical effect. That possibility, perhaps unsurprising to the outsider, is the essence of the molecular heresy that now confronts evolutionary theory based on living animals and fossils.

'The zoologists who have the best command of information on dates have maintained a reserved scepticism towards the entire protein endeavour.' This comment on a lack of collaboration from the orthodox was made by Richard Dickerson, who works with cytochrome c and other protein molecules at the California Institute of Technology. A biologist who studies the kinship and ancestral relationships of animals by close scrutiny of their intricate forms of life can hardly be pleased when chemical analyses by people who are, in his eyes, scarcely biologists arrive at much the same conclusions from a mere molecule. But professional pique is the least of the problems.

Much more troublesome is the evidence of long-term regularity in the rate of molecular changes. The molecules, or more precisely the genes which code for their manufacture, seem to be more or less indifferent to the great biological and environmental events in the history of life on Earth – even to the movements of continents or to the rise of wholly new kinds of plants or animals.

Changes in the cytochrome c molecule occur at a rate of about two altered chemical units every 100 million years, irrespective of whether the gene is being carried by mighty reptiles or cells of yeast. Since the time, 450 million years ago, when the most recent common ancestor of the pig and the tuna lived, the line of descent to the pig has accumulated about nine mutations, and fish's ancestors have also accumulated about nine, producing the seventeen chemical differences found in the cytochrome c of the present-day animals.

Pockets of conservatism

Haemoglobin, the vital molecule that carries oxygen in the blood, accumulates about eight mutations per 100 million years. A working molecule that changes faster still is fibrinopeptide; its function is very simple, as a spacer in the body's manufacture of another protein, fibrin, during the formation of blood-clots. These differences lead to the notion that some molecules are freer to evolve than others, because their precise composition is less important to the organisms that use them. Within each molecule, too, some portions change readily while other parts are more conservative; the most critical parts remain the same in every species. In fact, the parts that do not change identify themselves to the molecular biologist as the 'active sites' where the molecule carries out its duties. And when changes occur in marginally critical sectors of the molecule they involve the substitution of a chemical unit that is very similar to the one that it replaces.

That some mutations matter more than others becomes abundantly clear from ongoing mutations in human haemoglobin. For the species as a whole there is a 'standard' form of the molecule, and sickle-cell haemoglobin occurs frequently in malarial regions. But in addition 150 mutant forms of haemoglobin have been discovered so far in human beings; there must be other mutations still undiscovered. Of those known, many seem to do no harm whatever to the individuals; they are usually changes on the outside of the molecule. About twenty chemical units form the working zone of the molecule, clustered as a pocket around the 'haem', the chemical that actually takes hold of the oxygen. Changes to any of these twenty are much more likely to have serious consequences. One grave example is haemoglobin Hammersmith, in which a change in a single chemical unit allows the haem to fall right out of the molecular pocket.

Most of the mutations of haemoglobin and other working molecules which occur in each generation die out quickly. Over periods of millions of years, a very few mutations become the norm in particular populations of plants and animals and their descendants. The odds against that happening are very long indeed, but evidently not long enough to prevent it altogether. They are rare in any one kind of molecule. But given the many different types of molecules and the genes that code for them, in complex animals like man, the overall rate of molecular evolution must be faster than anyone would have guessed ten years ago.

Hereditary roulette

The Monte Carlo method and the use that researchers into evolution make of it symbolise the role of chance in the life game. A computer acts as a high-speed roulette wheel. It throws up a succession of random numbers which it then employs in tracing the outcome of a long sequence of chance events. The method is not exclusive to biology. In the design of seaports, for example, it can predict how often ships that arrive at random will have to queue for berths; it also assists the physicist in mastering uncertain processes in the sub-atomic world. For the biologist, the Monte Carlo method is a way of following the fate of genes.

The computer experiments trace evolution through many generations in imaginary populations of animals and plants. Suppose that fifty per cent of the animals in a particular population have a gene specifying brown whiskers, while the rest have another form of the same gene that makes the whiskers black. Suppose also that the colour of the whiskers makes no practical

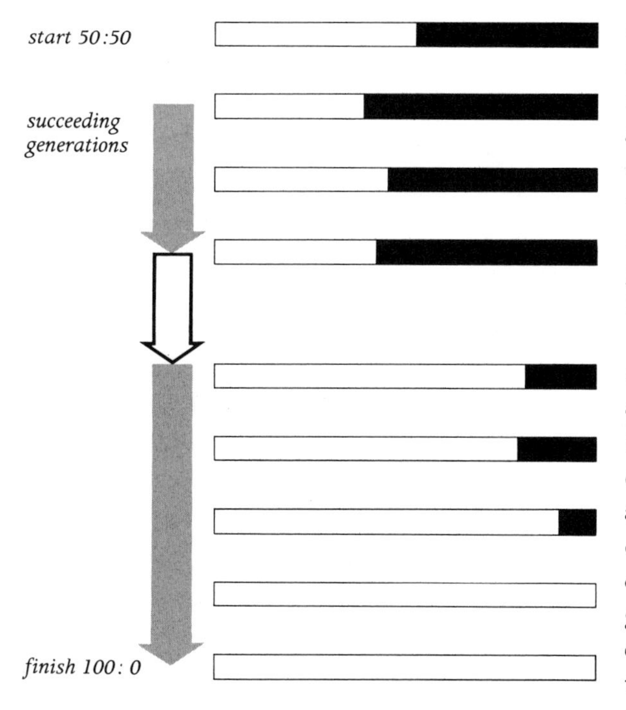

Evolution as a gamble. Even if two forms of a gene are at first equally common in a population, and neither is favoured or rejected by natural selection, sooner or later one will completely displace the other, simply by chance.

difference to the individuals that possess them and natural selection does not prefer one to the other. Even so, the simple intervention of chance prevents animals of the two kinds from leaving exactly the same number of offspring. For example, the next generation may be forty-seven per cent brown-whiskered and fifty-three per cent black-whiskered. Successive generations may fluctuate much more widely than that, from the initial 50:50 balance of whisker colours.

Monte Carlo experiments demonstrate processes of that sort. Being a matter of chance, the outcome of any one experiment with one imaginary population is uninstructive by itself. It could just as well be done with a real roulette wheel, or by drawing brown and black balls from a bag. The computer, though, can take a hundred imaginary populations and quickly follow their fate through many generations. It gives a speed of 'reproduction' and 'evolution' that even bacteria cannot match. Patterns of probability then begin to emerge as certain kinds of outcome turn out to be the rule rather than the exception. In our simple example the Monte Carlo experiments show that the whiskers of any population will infallibly finish up either all brown or all black. Purely by chance the population evolves, eventually 'fixing' one of the alternative genes and wiping out the other.

Even if only one animal in a population of 1000 possesses a peculiar gene, the same process operates. The odds are then massively against its survival. If the gene is harmless, though, it has one chance in 2000 (twice the population size) not merely of surviving but of completely displacing the normal form of the gene. Monte Carlo experiments bear out this theoretical prediction. The peculiar gene requires something like 4000 generations (four times the population size) to accomplish its takeover. Given plenty of mutations and plenty of generations, these inherently

Motoo Kimura, the leading heretic of modern evolution theory, asserts that chance plays a much larger part in evolution than most biologists like to think. Pictured here with him, in the computer room at Mishima, is his colleague Tomoko Ohta.

improbable events will actually occur quite often.

A Japanese population geneticist, Motoo Kimura, thinks that they are happening far more frequently than previous theories of evolution have allowed, and that chance must now be accorded a much greater part in evolution. Playing his Monte Carlo games at the Japanese National Institute of Genetics in Mishima, Kimura has emerged as the arch-heretic of modern evolution theory. Meeting him in Mishima, I did not find him typecast for this role. He is mild in manner and shy of fierce debate; he spends his spare time not pamphleteering but growing orchids. Kimura built up a solid orthodox reputation in the United States, where he helped to complete the very edifice of New Darwinism under which he was shortly to deposit a bomb.

In February 1968 Kimura published a short paper in the journal *Nature*. In it he reflected upon the new information about the remarkable differences among haemoglobin, cytochrome c and other molecules in various species of animals. Taking account of the large numbers of molecules and genes, in human beings for example, a mutation in one molecule or another must become newly but permanently established in the human population every few years. But according to calculations of the distinguished geneticist J. B. S. Haldane (who died a few years previously), the action of natural selection in substituting a new version of a gene for an old one requires a certain number of selective deaths of individuals. This number is so great that the human species could 'afford' to establish a new gene only once every 1000 years or so. The cost of faster evolution would be more than a species could bear.

Kimura reasoned that the high rate of molecular evolution made sense only if many mutations were nearly neutral to natural selection – being, in other words, neither harmful nor beneficial to the owner and exacting no 'price' in selective deaths for their introduction. The other recent discovery, that molecular alternatives were much commoner in living populations than most biologists supposed, also fitted Kimura's argument. They represented, according to him, nearly neutral mutations 'drifting' through the populations. This theory meant that, over a great deal of evolution, natural selection had no control. The ensuing controversy continues to this day. A brief excursion into history may help to put it into context.

The evolution of evolution

Jean-Baptiste Lamarck, the great French naturalist, recognised the fact of evolution before Darwin did; he died when Darwin was still an undergraduate at Cambridge, with his ideas having found little favour. His guess about how evolution occurred was that animals developed various organs as they used them to secure the necessities of life, and that changes so wrought in their bodies affected the inheritance passed on to their offspring. Given a genetic system different from the one we have (about which Lamarck knew nothing), life on another planet could presumably evolve in that fashion. The 'inheritance of acquired characters' is not as ridiculous an idea as it is often made out to be. Even Darwin flirted with Lamarckism to circumvent some difficulties in his main theory.

In casting around for the prime explanation of evolution Darwin was influenced, as were others, by Thomas Malthus' idea that premature death must be a normal feature of existence. Populations would tend to breed more offspring than the environment could support. But which individuals would then perish, and which survive? Reflecting on this question Darwin

The contrasting colours and patterns of ornamental Japanese carp are mutations accumulated in two centuries of breeding.

was led to the idea that the survivors would, by and large, be those better fitted to the environment. Darwin set out the principle of natural selection in evolution but he knew no better than Lamarck the source of the inheritable differences between individuals which gave nature the choices from which to select.

He was not aware of Gregor Mendel, busily cultivating peas in Moravia. Mendel's discovery of the gene as the unit of heredity, though one of the greatest of all accomplishments in biology, languished unnoticed in an obscure publication for more than three decades. After it came to light in 1900 a third kind of explanation for evolution arose. The early geneticists were impressed by the 'all-or-nothing' quality of hereditary characters and mutations represented in the genes that were convenient for study. Mendel's peas were either green or yellow, not yellowish-green or greenish-yellow. This contradicted Darwin's emphasis on gradual variability between individuals within a species, and it encouraged in the geneticists a different sense of how evolution worked.

There must be, they thought, a perfect type, an ideal set of genes for a given species. Variability between individuals represented unfortunate mutations – departures from the ideal. Evolutionary progress did not occur gradually but had to wait upon very rare favourable mutations. The realisation that many characters, such as stature or brain-power, were determined by many genes acting in concert softened this 'mutationist' version of evolution, but it still differed in principle from Darwin's. It gave chance, in the form of 'lucky' mutations, the starring role in the drama and left natural selection in the supporting cast, merely weeding out the 'bad' mutations.

A quarter of a century ago, the three main explanations of evolution were still in open conflict: Lamarckism, Darwin's selectionism and mutationism. Of these Lamarckism, the idea that experience directly modified heredity, could muster no hard evidence whatever in its support; it was kept alive only in Russia and Stalinist Europe, although a few biologists in other places continued to countenance the idea. The serious scientific battle was in the West, between the selectionists and the mutationists. They agreed that differences between individuals stemmed from mutations and that natural selection operated on these variations. But the selectionists held that variations were normal and healthy, while the mutationists thought they were aberrant and usually disadvantageous.

By the 1950s mutationism was in full retreat and the ideal type was seen to be a myth. Treatment of genetics in terms of populations rather than of individuals or types relegated mutations to the chorus of variable hereditary material available for natural selection to work upon. Strange to say, many descriptions of evolution, including Jacques Monod's *Chance and Necessity* (written in 1970), are still following the obsolete ideas of mutationism. Monod says, for example, 'all the properties of living beings are based on a fundamental mechanism of molecular invariance' and for him variations are 'imperfections' or 'noise'. From this he has deduced that 'evolution is not a property of living beings'. To modern evolutionists Monod's account of life and evolution seems to start from a false premise: inherent variability, not invariance, is the essential ingredient of living things.

The geneticists came back to the fold of selectionism and united with other evolutionists in constructing the New Darwinism, which by 1960 was substantially complete in its present form. Within this framework of ideas there was only one misfit.

A colony of Peruvian cormorants. In populations of animals alternative forms of genes and the molecules they make become commoner or more rare during successive generations. That is the very essence of evolution.

It was the notion of 'genetic drift' which suggested that the luck of reproduction, especially in local populations, could greatly help in establishing favourable combinations of genes by random fluctuations; in other words chance could sometimes negate, sometimes assist natural selection. But this was a minor theory that found little support. Darwin's great principle was fully restored among most biologists.

Virtually everything that happened in the life game was now to be understood as the work of natural selection, which appraised every genetic mutation as favourable or unfavourable to the organisms. Any difference in a gene's performance, however slight, would eventually be found out. There was no room for doubt. By 1969 Ernst Mayr of Harvard University, widely regarded as chief warden of the New Darwinism, felt able to write of modern selectionist theory: 'It is the application of the theory that is sometimes controversial, not the theory itself'. Mayr already knew of Kimura's bomb of course, but then, as now, he thought it could do no damage.

Too many mutations

The emergent science of molecular biology at first suited the New Darwinists very well. All that was being done to reconcile genetics and natural selection was lit up by the discovery of the chemical nature of the genes. Moreover, the genetic machinery turned out to be of a kind that disposed once and for all of the worst heresy – Lamarck's. The so-called 'central dogma' of molecular biology, well attested by all the available evidence, is that while a gene can make a protein, and a mutant gene a modified protein, the character of a protein cannot be communicated back to the genes. In other words, genetics at a molecular level is a one-way street. Effects of the environment which alter the outward character of the plant or animal can no more alter that organism's genes in a coherent way than crashing a motor car could magically affect the presses at the factory where it was made.

Molecular biology promised another benefit. Enzymes and other working molecules are chemical translations of the genes, and in principle at least any genes could be examined by fishing out the molecules for which they coded, and comparing them. Before then geneticists had been decidely limited in their choice of genes whose effect they could study directly. Eye colour, the number of bristles on a Drosophila's body, blatant genetic disease, the indirect classification of molecular differences as in the case of blood groups – only conspicuous features such as these were easily accessible to them. Only the vaguest guesses could be made about important questions such as how many genes were involved in changing one species into another. Yet genetic theories were available for dealing with large numbers of genes in large populations. As one expert put it, there were 'theories in search of facts'.

The facts that came, in the form of new information about molecules, were more than the geneticists had bargained for. The diversity of genes performing the same function, discovered by molecular techniques in the mid-1960s, provoked the present crisis in evolutionary theory. First there was the unexpected abundance of the molecular alternatives in living populations; then there were the apparently gratuitous and continuous changes through evolutionary history, revealed by comparing equivalent molecules in different species. Many molecular biologists began to take it for granted that in many instances natural selection could have no preference for this version of a molecule or that. Motoo Kimura was the first evolutionary theorist to set out the full implications, as he saw them, of molecular diversity.

As we left the account earlier, Kimura was broaching, in 1968, the idea that many mutations must be nearly neutral in their effect, otherwise the high rate of change in working molecules would be lethal. This idea is quite different from the old mutationism, the essence of which was that every mutation mattered greatly. Many biologists, though, have likened Kimura's theory to the proposition, also mentioned before, about 'genetic drift' in small populations in special circumstances. The new theory also implied random genetic drift – evolution governed by chance rather than by natural selection – but on a much grander scale than in the earlier theory. For Kimura, in all organisms past and present, most evolutionary changes have escaped the control of natural selection. For what he describes, there is no real precedent in evolutionary theory.

Putting his ideas in their simplest form, Kimura compares his own molecular constitution with a carp's:

The surprising fact is that most of the mutations do nothing to help establish the differences between human being and fish. The carp and I both need haemoglobin to do exactly the same job of carrying oxygen around the body. Yet one-half of all the chemical units in my haemoglobin molecules are different from the carp's. That unnecessary sort of evolution, and my studies of its rate and pattern, suggest to me that natural selection has had no reason for preferring one variant of the molecule over another.

According to the selectionist theory nearly all mutations are harmful. A very few beneficial mutations occur and may be adopted in the population. A very few neutral mutations may occur but they tend to be lost.

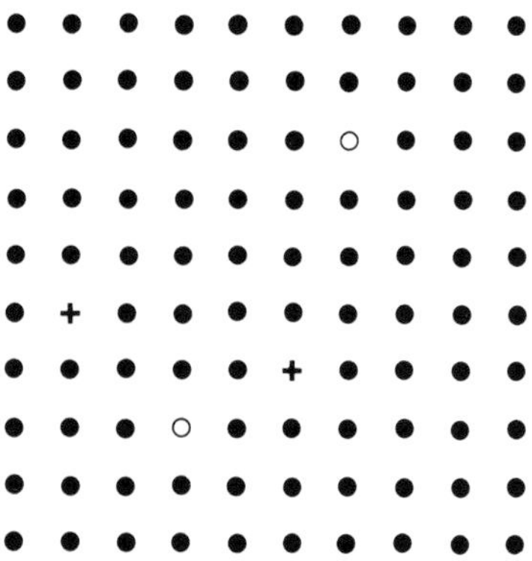

● *harmful mutations*
+ *beneficial mutations*
○ *neutral mutations*

According to the neutralist theory, too, nearly all mutations are harmful. Even fewer beneficial mutations occur and most evolutionary change is due to neutral mutations which are quite common.

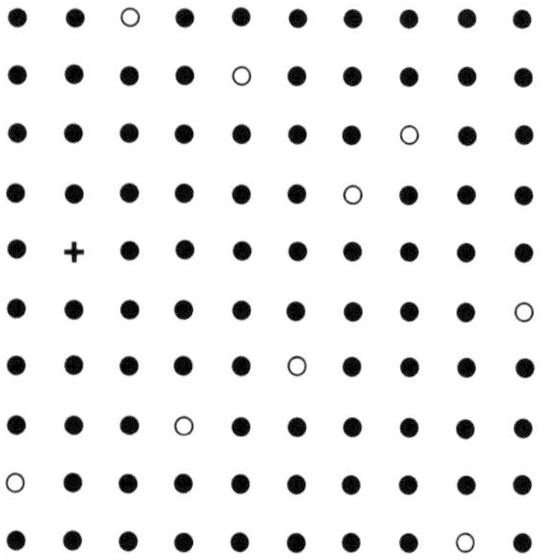

Symptoms of randomness

In 1969, in the journal *Science,* under the title 'Non-Darwinian Evolution', outspoken support for Kimura came from two scientists at the University of California at Berkeley, Jack King and Thomas Jukes. King is a young geneticist who has since moved to Santa

Barbara. Jukes investigates the origin and evolution of the molecules of life at Berkeley's space science laboratory; he had argued for a very high rate of molecular evolution even before Kimura launched his theory.

The nature of the genetic code gave King and Jukes a fresh argument about mutations and evolution in molecules. In the code, a word of three 'letters' – three chemical units of nucleic acid – specifies the incorporation of an amino acid into a working molecule. But more than one word can have the same meaning; some amino acids have six words for them, others only one or two. Examining the code in detail one can see what new words are made by random changes in existing words, and what amino acids are thereby specified. Sometimes a change in the word leaves the amino acid unchanged. For example, GGA can change into GGC and still stand for glycine; but a change from GGA into GCA alters the amino acid to alanine.

All this leads to a prediction of how often each kind of amino acid will turn up, given purely random mutations. For instance, glycine should be twice as common as tyrosine. The pattern of occurrences in actual working molecules of animals comes very close to the predicted pattern. This result makes little sense if natural selection is exerting 'editorial' control, because the merits or otherwise of a particular gene ought to have nothing to do with what words it uses, any more than a poet should select his words according to how easily they can be made by misprints of other words.

Some changes from one amino acid to another require two alterations to the word in the genetic code. Yet even these changes occur remarkably frequently – in more than a quarter of the changes detected in each phase of molecular evolution. If, as seems likely in most cases, the necessary mutations do not occur at the same time, then the selectionists require that first one mutation and then the other should have advantages in their time, and be selected. Walter Fitch of the University of Wisconsin concludes that, if natural selection were fully at work, these 'double mutations' should be a good deal less common than they are.

Fitch has been involved in molecular evolution for a long time. With Emanuel Margoliash he was responsible for much of the work in the 1960s which established the basic patterns of molecular evolution and the approximate constancy of the rates of change. He has also reasoned that only a minority of units in each molecule are free to change at any one time, even in the variable portions of the molecule. When this limitation is taken into account, the proportional rates of change in different molecules – cytochrome c and haemoglobin for example – become much more alike.

But Fitch sees that the true constancy or otherwise of those rates of change, in different phases of evolution, may be important for deciding whether the changes are mainly neutral or subject to control by selection. In a renewed inquiry into this question, Charles Langley, a young Wisconsin colleague, and Fitch have pooled the information about changes in several different molecules, in twenty-four different species of animals from fish to man. That gives them better statistics with which to work. They confirm that the rates of molecular evolution vary quite markedly. In particular, molecules in apes and man have been evolving slowly, in rodents rapidly. Fitch comments:

> That molecular clock may be keeping accurate time over long intervals but those ticks are not one monotonous lockstep. Therefore we don't need a neutral mutation theory to explain uniformity, because there is no uniformity.

Kimura's closest ally has been his colleague at Mishima, Tomoko Ohta. She is a notable population geneticist in her own right and together Kimura and

Ohta have sharpened and refined the theory of nearly neutral mutations. In their more recent presentations of it, some salient points appear. One is a remarkably simple mathematical conclusion. The rate at which new neutral mutations establish themselves in the population is exactly equal to the rate at which neutral mutations appear in each sperm or egg that goes into making a new individual.

Suppose everyone in a population of a thousand has twenty new neutral genes, ten from the sperm and ten from the egg. In due course ten out of the 20,000 new genes will become widespread in the population; 19,990 are lost. But the answer is still ten even if the population is only a hundred, because new genes insinuate themselves more easily into small populations. New neutral genes will thus 'fix' themselves at a constant rate regardless of population size and give the more or less steady rate of molecular evolution. Uncertainty remains about whether the rate should be reckoned in years, allowing for the time over which new mutations accumulate in the sex cells, or in generations, marking the intervals at which sperm and egg come together. Over periods of millions of years, new forms of genes establish themselves in the population and then are themselves gradually replaced. This turnover of genes, governed purely by chance, is responsible for the coexistence of so many molecular alternatives.

In response to his critics, Kimura concedes that some of the changes in molecules through evolutionary history may have been useful to the species involved and their presence therefore reflects a positive influence of natural selection. The same may be true of some of the molecular alternatives in living populations. But he insists that the great majority of these changes have no biological meaning. Kimura and the selectionists are agreed, though, that nearly all new mutations do have biological meaning – the mutations are harmful and are eliminated by natural selection. Both sides accept also that the appearance of a beneficial mutation is a rare but evolutionarily very important event, with natural selection again coming into play to improve its prospects for incorporation into the population.

The essential point of difference is as follows. The selectionists suppose that every gene that is common in the population has been carefully chosen by natural selection for positive qualities it possesses. Kimura and the 'neutralists' assert that, on the contrary, every gene incorporates random changes about which natural selection says nothing.

Darwinian backlash

Dr Kimura and his followers say that evolutionary changes are neither useful nor harmful to their possessors. . . . If that were so, evolution would have hardly any meaning and it would not be going anywhere in particular. All that we know, both from laboratory experiments and observations in nature, shows that evolutionary changes enhance the chances of the organism to survive and perpetuate itself in some environment. And this is not merely a quibble among specialists. To a man trying to understand the meaning of his existence, evolution by natural selection makes sense. It is an unconsciously creative process gradually and slowly improving the organism, making it better adapted to its environment.

Theodosius Dobzhansky, who offers this rebuke to Kimura's ideas, is a monumental figure in evolution studies. In his career he has combined the predilection for theory of his parental Russia with the vigorous experimentation and field work of his adopted America. Now in his seventies, and based at the University of California at Davis, where I met him to discuss the theory of neutral genes, Dobzhansky

Theodosius Dobzhansky, in his Californian laboratory.

continues to interrogate his beloved Drosophila for the evolutionary secrets they embody. He still travels to remote jungles and mountains to hunt the flies. Earlier he played a leading part in bringing about the grand reconciliation of Mendel's genetics with Darwin's natural selection. But the resulting 'selectionist' genetics is precisely what Kimura is challenging. To emphasise his dislike of the new heresy, Dobzhansky has even compared Kimura with Lamarck – leaving the Japanese theorist uncertain whether to be offended or flattered.

Ernst Mayr of Harvard University is another outstanding evolutionary theorist who has little time for Kimura or his allies. 'A random replacement of amino acids unquestionably occurs occasionally in evolution,' he comments, 'but it appears at present that it does not anywhere near approach selection in importance as an evolutionary factor.'

On the whole, molecular biologists are inclined to think Kimura may be right, while biologists working with living animals are for the most part opposed to him. But scientific issues have to be settled by facts, not by opinion polls or the heavy hand of authority. So many theoretical and experimental studies are now dedicated to proving Kimura wrong, that I can only give examples.

One approach is to see if Kimura's theory leads to absurd conclusions. John Maynard Smith, whose use of games theory in evolution figured in the previous chapter, has argued that more than one 'normal' version of haemoglobin should exist in the human population, if Kimura were right. Abnormal forms abound, of course, but the only reasonably common alternative forms of haemoglobin are those, such as the sickle-cell type, which are plainly associated with malaria and hence natural selection. Either Kimura is wrong, Maynard Smith reasons, or else the human population went down, at some stage, to such a small number that the neutral variants of haemoglobin were eliminated by chance.

Tongue-in-cheek, he offered Adam and Eve as a possible bottleneck of the required kind, but other biologists suggested Noah and his family. An excellent idea! Maynard Smith and his Sussex colleague, John Haigh, promptly agreed that the biblical story of Noah provided the best test of Kimura's theory at present available. As 'unclean' animals (pigs, camels, etc.) were admitted only as single pairs into the Ark, compared with seven pairs of the other animals, those unclean animals should show fewer variations among their genes. I have not heard of anyone launching a research project to examine the appropriate animals.

An ardent selectionist prominent among the younger generation of experimental geneticists is Francisco Ayala of the University of California at Davis. Tests to show that natural selection is not indifferent to the variants of genes provide him with his main attack on the Kimura theory. In one recent experiment he began rearing Drosophila flies from chosen parents among which a particular version of a gene was much scarcer than in the same species in the wild. Over the succeeding months the gene rose in frequency with every generation, steadily approaching the norm. Ayala concludes that the preferences of natural selection are here expressing themselves very clearly; if Kimura were right the frequency of the gene should go up or down at random.

Other investigators are at pains to show, in detail, how particular variants of molecules benefit a species. The rainbow trout, for instance, has two versions of a vital brain enzyme, acetylcholinesterase. Far from being redundant, one of them functions in cold water and the other in warm water. And tackling the molecular evolutionists on their own ground, a group at the State University of New York in Buffalo has studied the fastest evolving molecule known – pancreatic ribonuclease. They interpret the changes, even in highly variable and apparently non-critical parts of the molecule, as being governed by the needs of molecular architecture and hence susceptible to natural selection.

Against one of the strongest features of Kimura's theory, the way it accounts very simply for the roughly steady rate of change in molecules during evolutionary history, the selectionists have two courses open. They can explain the constancy in their own manner. They say that such continual opportunity exists for working molecules to become better matched one to another that their rate of evolution is limited only by the rate of mutation, which is roughly constant. The other possibility is to show that the constancy of the rate of molecular evolution is an illusion or at best a very crude average, concealing marked variations that occur in particular evolutionary lines of descent at particular times. But the variations in rate known so far are not very damaging to the neutralists' cause.

An indelible mark

My reason for perplexing the reader with some examples of the contentions on both sides is to convey a sense of the current impasse. The theoretical and experimental efforts so far have done very well in confirming both the selectionist theory to the satisfaction of the selectionists and the neutral mutations theory for the convinced neutralist. This phase of a major dispute, when different experts make irreconcilable statements with equal vehemence and equally plausible arguments and evidence, I have experienced in other branches of science. I have seen David slay Goliath, or, in other contexts, Goliath trampling on

David. Either way the challenge to the giants of orthodox thinking is a splendid stimulus to new advances in science. Resolution in the present case will come, maybe weeks, maybe decades from now, when the selectionists have evidence that persuades the neutralists they were wrong, or vice versa.

Some biologists try to resolve the controversy by compromise. First they minimise it. Everyone agrees, they say, that natural selection works on all mutations except a minority of neutral ones; the only dispute is about how many neutral ones there are. Then they suggest that nearly neutral mutations are more common than the selectionists thought but not as frequent as some of Kimura's estimates. Yet such an outcome would be very much to the advantage of the neutralists. The contest is not symmetrical, because until Kimura came along the selectionists claimed they had the basic answers and that neutral mutations were insignificant.

Others try to dismiss the controversy by saying it does not matter: neutral genes by definition make no practical difference to the plants or animals that carry them so they have nothing important to tell us about evolution. How much less interesting, they remark, than the evolution by natural selection of the bird's wing and the man's brain. But it is only invisibility of molecules that allows anyone to reason that way. Freckles are roughly neutral mutations occurring in the cells of the skin during life; they are small and round because only local populations of cells are affected. If hereditary neutral mutations also produced variously coloured spots on our skins we should not be able to dismiss their contribution to the diversity of life.

In any case, the potential evolutionary significance of neutral mutations is much greater than that. Changes in molecules are the source of all minor and major steps in evolution. The environment of organisms is changing all the time and the organisms are evolving all the time. A gene that is neutral today may, in new circumstances a thousand generations hence, come to assume great significance, benign or harmful. If the gene were not tolerated today, it could not survive to exert its effect in the future. Again, if trivial changes can accumulate in a molecule over many millions of years, without interference from natural selection, the day may come when a peculiar molecule thus produced goes through one more mutation that suddenly gives it new importance – just as changing one card can alter a worthless poker hand into a royal straight flush.

The neutralists' chancy theory of evolution can also be extended into the selectionists' own territory. By its emphasis on random effects it casts light on the conditions in which beneficial new mutations are most likely to succeed, if they appear. In particular Kimura's colleague Tomoko Ohta has offered a 'neutralist' explanation of why selective evolution is fastest in small groups of organisms in a specialised environment. Such groups are highly 'experimental', risking extinction but also having more chance of establishing valuable new genes. In a large inter-breeding species, occupying various and changeable habitats, an evolutionary innovation has much less chance of becoming fixed. This is somewhat different from the older 'genetic drift' theory mentioned earlier but, like it, Ohta's theory gives chance a much bigger part, even in evolution by natural selection, than the selectionists would like to think .

To leave the New Darwinism of 1960 intact only a complete rout of Kimura and his fellow heretics will do. Such an outcome seems less likely with every year that passes. Everyone can pick among the evidence for the pieces he likes, but Walter Fitch's work

Fossil colonies of primitive microscopic plants. Such 'stromatolites', composed of vast numbers of blue-green algae, prosper today in salty water at Shark Bay in Western Australia. They represent a very ancient form of life.

on the double mutation, for example, is persuasive evidence that quite a few, at least, of the mutations are indeed neutral. Closely related to that is the question posed by Brian Hartley, with whom we began the chapter: where do new enzymes come from? Too little attention has yet been paid to the argument that accumulating the multiple mutations needed, if a working molecule is to change its function, demands long-lasting 'silent evolution'. Hartley and his colleagues, though, do not side with Kimura; they think that natural selection is eased but not lifted entirely.

Whatever the outcome, molecular biology has already made an indelible mark on studies of evolution. However they are to be explained, the abundant variations in genes, revealed within species and between species, provide a novel, surprisingly busy picture of the life game in progress at the molecular level. It supplies the geneticist, at last, with plenty of facts to work with. And molecular biology brings completely fresh insight into the past evolution of life on Earth, not merely for comparison with the fossil record, but for reaching back to crucial stages of evolution for which no fossils exist.

Even so, the issue of neutral mutations remains wide open. Richard Lewontin helped to set the stage for the dispute, with the discovery of the abundant molecular alternatives. He, too, has been active in recent research bearing on the issues. But he has been as puzzled as anyone, to tell whether the neutralists or the selectionists are more nearly right. He says:

What's really at issue is the leading problem of evolutionary biology today. Is Dobzhansky right, that most of the genetical variation we see in natural populations is a result of natural selection and will serve as a basis for future evolution in those species, for future adaptation? Or is Kimura right that most of that variation, if not all of it, is neutral and has nothing to do with evolutionary adaptation. . . ? I don't think that . . . any experiment . . . or any theoretical development that I can imagine today will really decide the issue. . . . Although I do think that Kimura has done us a great service in calling attention to the importance of random events, I myself have no great passion about the issue and I don't come down on one side or the other.

As we turn next to what Darwin called 'the far higher problem of the essence or origin of life', the question will still dog us, about the parts played by chance and natural selection in the life game. But in current research dealing with events when the Earth was young, we shall find a curious reversal. In this chapter, molecular biologists have been invoking chance where selection was supposed to rule; in the next, they summon natural selection into a domain hitherto reserved for chance alone.

Chapter 4 Grandmother Hypercycle

Mars has a barren look. The views of that planet's surface transmitted by the American *Mariner 9*, and information returned by instruments in various other spacecraft, are discouraging for seekers of life. The volcanic surface is dusty and cool, with little trace of the water that is essential for life as we know it. Yet the possibility remains that on Mars we shall shortly encounter the first living organism outside our own planet and uncover a whole new story of evolution where the life game has been played in circumstances quite different from our own. That would be the greatest discovery that can be expected from space exploration.

The search proceeds on the assumption that life on Mars, if it exists at all, will resemble life on Earth in basic respects – such as dependence on carbon. Perhaps the human species is lacking in imagination, but to envisage anything that we would call living is very difficult except by resort to the wonderfully elaborate chemical compounds that carbon is capable of making. No other chemical element shares its ability to form long chain-like molecules, with chemical bonds between carbon and carbon and between carbon and other elements, which are neither too strong nor too weak. Some people have suggested 'living rocks' based on silicon. For the planets Mercury and Venus where carbon compounds would char, or the cold planets like Saturn where they would freeze, a radical possibility like this may be the only hope for life. But it is a slender one indeed. On Mars the climate appears tolerable at least for hardy forms of carbon-based life and there is carbon in plenty in the atmosphere.

At the Jet Propulsion Laboratory of the California Institute of Technology, one of several instruments, intended to go to the surface of Mars to test for the presence of life, is being prepared for *Viking* spacecraft. It will expose a sample of the Martian soil to light and to radioactive carbon gases; if there are any microbes resembling plants they will assimilate radioactivity. In laboratory tests, using farm soil and the like, the instrument has successfully confirmed the presence of life on Earth. But Norman Horowitz and his colleagues, who designed the instrument, have also been carrying out experiments to see how readily or otherwise the Martian atmosphere would create chemicals of the kind needed for life.

They shine ultra-violet rays, as if from the Sun, on small containers enclosing carbon dioxide gas with traces of carbon monoxide and water vapour, together with various minerals. Subsequent chemical analysis shows that several promising chemical compounds, including sugar-like materials, ought to be forming spontaneously in the present-day atmosphere of Mars. Even if no living things are to be found on Mars, some chemical prerequisites for life have probably accumulated there in abundance. These experiments are very similar to others which reconstruct the conditions on the early Earth, in attempts to deduce how life began on our own planet.

The exploration of Mars bears closely on the problem of the origin of life on Earth, and may afford the only opportunity for observing, rather than speculating about, events of the kind that occurred billions of years ago on Earth. Before life itself altered the chemistry of the atmosphere, the Earth was more like present-day Mars than present-day Earth. The principal gases were probably ones that we should regard as suffocating or poisonous: carbon dioxide and carbon monoxide. Oxygen was lacking but key

A molecule game simulates the formation of the first genes, for the origin of life. If one of the chemical units, G, mutated into a C it could form a cross-link with the adjacent G and make the molecule more stable.

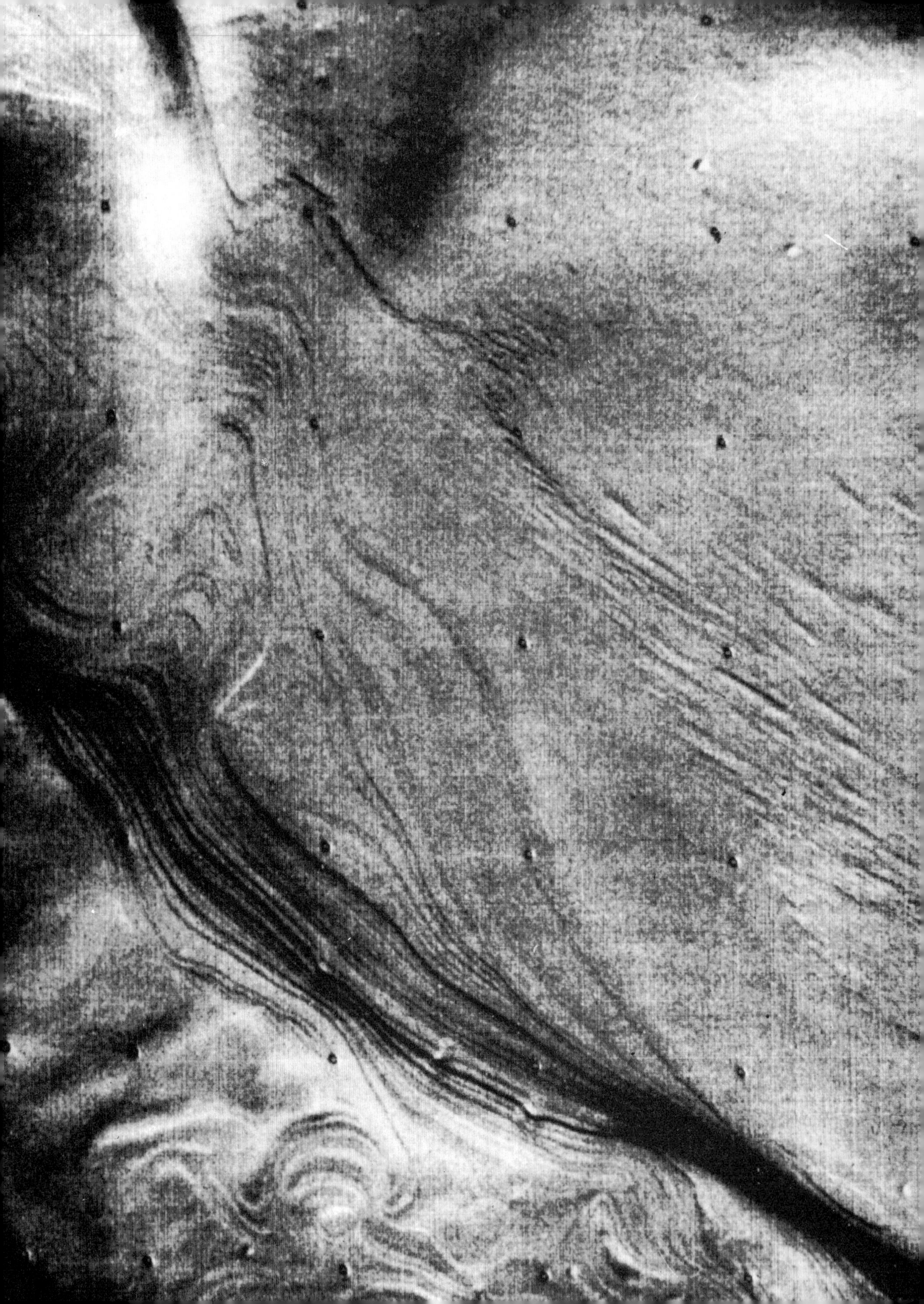

The surface of Mars, near the south pole, televised by the Mariner 9 *spacecraft. Mars is the only likely candidate among the other planets for possessing life, and its exploration is closely relevant to the question of how life began on Earth.*

elements such as phosphorus were available, while ammonia may have existed as a valuable source of nitrogen. The Earth had much more water than Mars has today, and it was in water that life began.

Recipe for soup

For twenty years Stanley Miller, starting at the University of Chicago and now working at the University of California at San Diego, has been testing mixtures of materials that may have existed at the surface of the young Earth. He supplies energy in the form of an electric discharge and sees what chemical compounds would have formed spontaneously. Other chemists have followed his example and today there is a most impressive list of carbon-based compounds, including chemical units needed in nucleic acids, proteins and sugars. These must inevitably have formed in great quantities if present ideas about the composition of the Earth's early atmosphere are even approximately correct.

Ultra-violet rays from the Sun were the most abundant source of energy for chemical reactions, according to another leading investigator of the origin of life, Leslie Orgel of the Salk Institute in San Diego. Assistance came from electrical discharges such as lightning strokes, from the shock waves of thunderclaps and meteorites, from radioactivity, from volcanoes and from cosmic rays. Orgel has estimated the production of chemicals by just one of these sources of energy – electrical discharges. Even assuming a much lower efficiency than occurs in laboratory experiments, it would be sufficient to coat the Earth to a depth of three feet in carbon-compound solids in the course of a few million years. Most of these materials would have accumulated in lakes and seas.

Such was the origin of the chemical soup. Almost everyone who considers how life began on Earth now assumes water rich in carbon compounds to be the source of the great event. Orgel supposes that the soup was about a third as strong, in its concentration of organic material, as Knorr's Chicken Bouillon. We can imagine an endless succession of chemical reactions going on between the many different kinds of materials occurring in the soup, as molecules were being built up and broken down by purely chance events. Any thickening of the soup, whether in volcanic lakes, in pools or tidepools, or even by freezing, greatly improved the chances of forming the long-chain molecules, nucleic acids and proteins, that would be indispensable for life.

Particular chemical aids helped to enrich the soup. One very simple compound, urea, promotes the incorporation of phosphates into organic compounds, an important step in the formation of nucleic acids and of other active organic molecules now used in living systems. Another, hydrogen cyanide, would encourage protein to form from its units, the amino acids. A common mineral of clay, montmorillonite, is capable of absorbing amino acids and converting them with almost complete efficiency into protein-like chains. In principle the soup, or some concentrated extract of it, could spontaneously produce almost any material found in simple living organisms.

Is any trace left of the soup, from before the start of life? The Earth was formed 4600 million years ago; the first unequivocal evidence of living things comes from rocks 3200 million years old. Older rocks, where soupy remains might be found, are very scarce. In any case the chemicals of the soup were not eternally stable; nothing would be left of them except petroleum-like or coal-like materials. Then the problem is to distinguish these materials from the similar remains of living organisms. Some

South African rocks about 3500 million years old contain carbon compounds. These compounds possess relatively more carbon atoms of a heavy form (carbon-13) than do carbon compounds left by the living materials. They may represent the lifeless soup. In slightly more recent rocks there seems to be a random collection of hydrocarbons mixed with a biased set of compounds typical of living things. But this is slender evidence indeed for the great transition from soup to life.

Clearer support for the soup theory comes from meteorites, the stones that fall to the Earth from space and may include pieces broken from another planet. A number contain carbon compounds of the kinds needed for life to begin, including a wide variety of amino acids. But analysts can quite easily show that the materials are not themselves products of life. Ordinary chemistry creates, in equal numbers, alternative forms of amino acids. They are 'left-handed' and 'right-handed', the one a mirror image of the other. But life on Earth employs the 'left-handed' kind of each amino acid exclusively. Any life employing complex carbon compounds probably has to be equally choosy, if the elaborate sculpture of its working molecules is not to be wrecked. Life elsewhere, though, could prefer the 'right-handed' amino acids. The amino acids in the meteorites are 'left-handed' and 'right-handed' in equal proportions. They are therefore genuine soup from somewhere else in the solar system.

Dice and cloverleaves

A great deal of time elapsed between the origin of the Earth and the first appearance of recognisable organisms – more than 1000 million years. Yet that was not nearly long enough for exactly the right materials to have gathered together by the unaided chance of chemical reactions. The simplest living thing requires the coexistence of many elaborate molecules, which are sufficiently well matched to co-operate chemically. The essence of life is information, embodied in molecules that make other molecules, which work in a required fashion. It is the enormous information capacity of biological molecules that makes their production by chance so improbable.

Life had no chance, if the chemical units of nucleic acids and proteins were simply combining and breaking up and recombining in a wholly random fashion in the soup. Any material you care to name could in principle be produced by the soup. But once you had named it, as a nucleic acid or protein with a particular sequence of chemical units, you would have to wait a very long time indeed for it to turn up. In living systems the selection, positioning and fixing of one unit into a molecular chain takes less than a thousandth of a second. Yet all the seconds that have elapsed since the formation of the Earth allow far too little time for purely random process to make a single practical protein.

It would be like trying to win the jackpot on an elaborate slot machine in which a particular sequence of symbols had to be lined up on hundreds of wheels revolving independently. Daunting reasons like these lead some molecular biologists to toy with the idea that life on Earth arrived from somewhere else, perhaps even being sown here by a technologically advanced civilisation. Other scientists say that the laws of physics and chemistry will have to be revised before 'the problem of life' can be resolved. The 'problem' is to bring the origin of life out of the wholly improbable into the realm of rare events that become more or less inevitable, given enough time. A German scientist, Manfred Eigen, thinks

The Eigen-Winkler game for making molecules. Each player starts with a random sequence of the chemical units (A,U,G,C) that compose a nucleic-acid chain. By throwing a four-sided dice he can 'mutate' selected units one at a time, trying to form the cross-links (A to U, or C to G) that would occur naturally in molecules like these. The realism of the molecules produced in the game is shown by the similar pattern of loops in the real nucleic-acid molecule in the diagram.

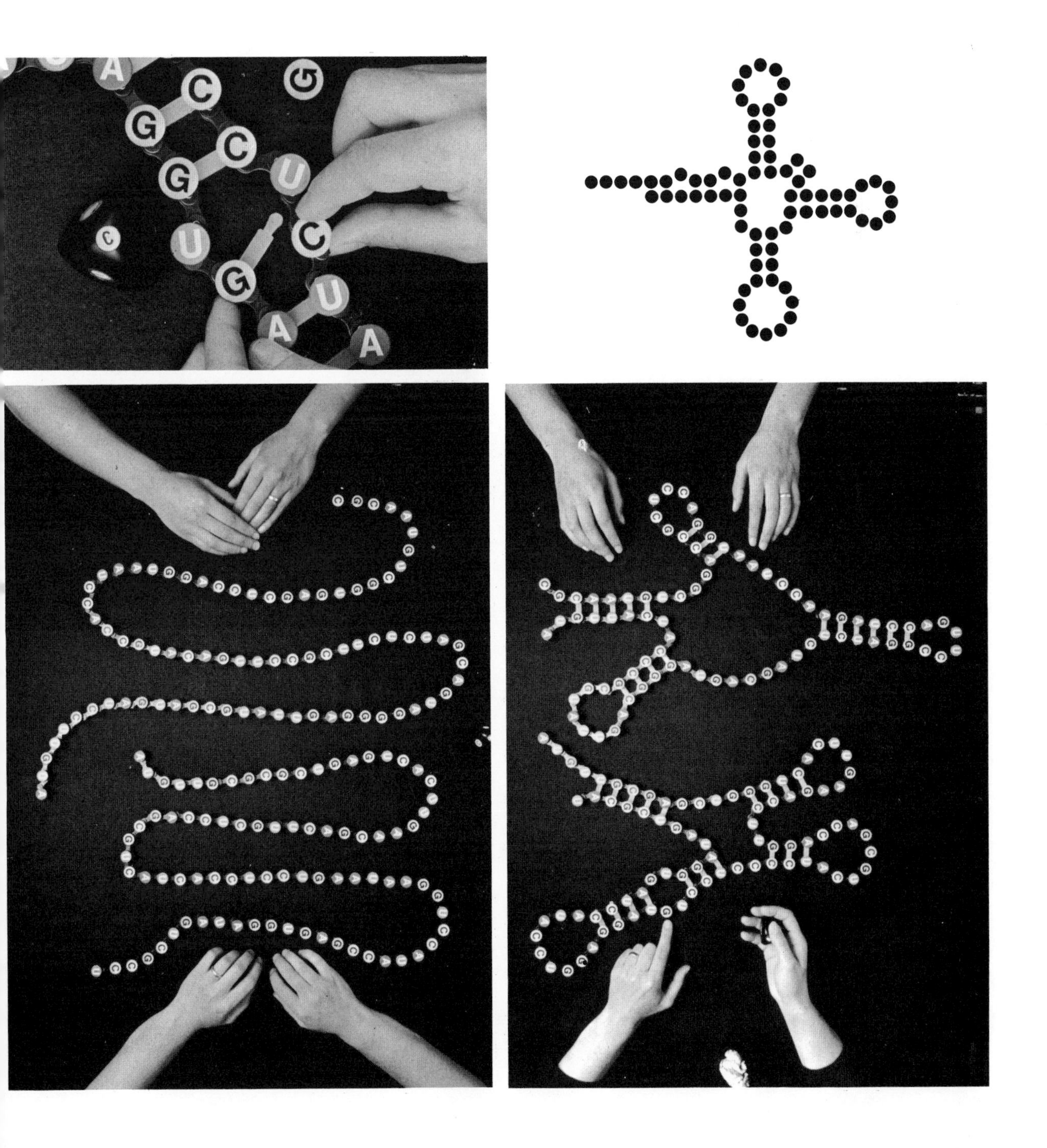

that a simple principle – an additional rule in the chemical game – overcame the dreadful odds against life. It was evolution by natural selection, starting long before life began.

As a chemist who won his Nobel Prize for finding out just how rapidly chemical reactions proceed among the working molecules of life, Eigen was at pains to emphasise to me the difficulties of assembling a particular sequence of chemical units simply by chance in the time available. He directs the Max-Planck Institute for Biophysical Chemistry in Göttingen, where he and his colleague Ruthild Winkler devised a game that shows how chance was given a helping hand in chemical evolution before life began. One question they tackle is: how did molecules with the right kinds of properties form and (just as important) survive in the chaos of the soup? Another is: what would have been the simplest combination of molecules that would secure, or at least compete for, its own survival, and evolve into the first fully-fledged organism? These two questions are closely linked.

Their game is a real one, though played for serious purposes. Tokens, normally 'popper' beads, fit together and represent the different chemical units, denoted A,U, G and C, that go into a nucleic-acid chain. Each player starts with a random sequence of these units. He then seeks to improve it by changing selected units; he does so in accordance with the throw of a four-sided die, which shows A,U,G, or C.

The crucial point is that the player can make a loop in his molecule so as to put an A beside a U or a C beside a G. It is chemically sound: these pairs of units normally combine quite readily when they lie side by side, each in its own portion of chain but now forming a cross-link. The rules are based on chemical measurements made at Göttingen. You have to make more than one cross-link at once, with C–G links

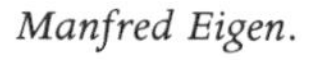
Manfred Eigen.

counting for twice as much as A–U links. By a combination of skill and luck each player tries to accumulate lengths of chemical units that can be laid side by side in this fashion, producing as many cross-links as possible.

The perfect loop, a hairpin, is not suited to trying out many combinations. Instead each player finishes up with a molecule with a number of loops radiating from it, like a cloverleaf. Clear proof of just how realistic the game is comes from the structures of actual nucleic-acid chains, notably the 'transfer-RNA' molecules which take part in the manufacture of proteins in real living cells. These too are organised as cloverleaves, or similar structures folded up, and the game shows why they adopt this form: structural advantages come from making loops.

In a real molecule the effects of the cloverleaf form are to stabilise it, to give it a definite shape and to protect it against damage or other changes. In the chemical soup of the early Earth, as in the Göttingen game, nucleic acids with appropriate structural qualities would have been more stable than others. Chance is already modified. But, for natural selection to come into play, molecules must be reproducing themselves. Only then can distinctions become apparent between molecules that survive to reproduce and those that fail to do so.

Reproduction is a comparatively simple matter for nucleic acids. In principle all that is necessary is for the loops in the chain to open out and for new chemical units to attach themselves, A to U and C to G, thus making a complementary version of the original molecule. When the process is repeated, a molecule identical with the original is produced. Provided no misprints have occurred the cycle is complete and can go on making the same nucleic acid molecules indefinitely. This ready replication is the basis of their

Sol Spiegelman.

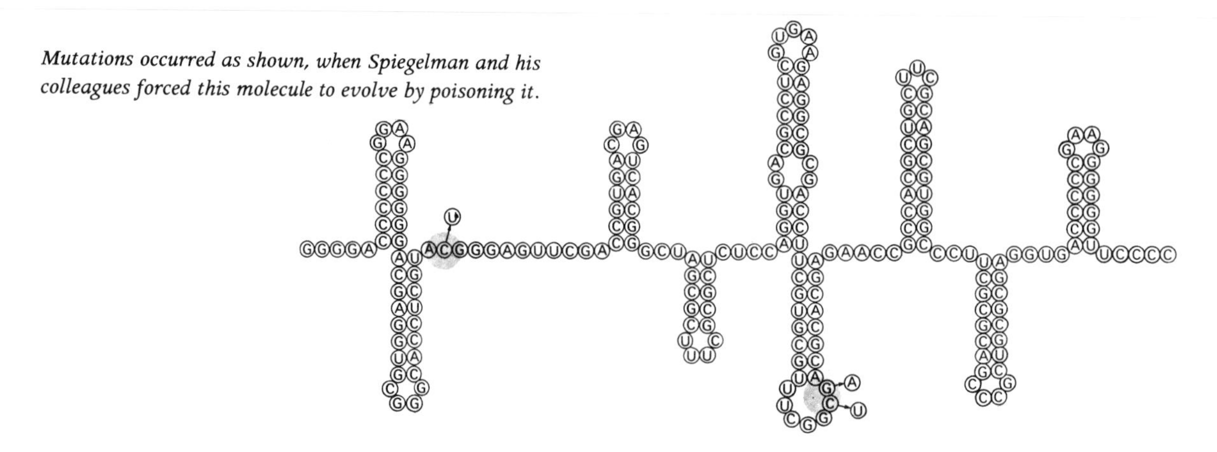

Mutations occurred as shown, when Spiegelman and his colleagues forced this molecule to evolve by poisoning it.

use in all organisms as the carriers of genetic information. In practice other working molecules help in replication; they give speed and fidelity. Without their help, reproduction of nucleic acids would have proceeded in the soup, but very slowly. Only very short sequences would have evolved.

Before following Manfred Eigen to the next, most elaborate step of his argument, let us briefly leave Göttingen for New York, to see the evolutionary potential of a piece of nucleic acid.

An adaptable molecule

Sol Spiegelman, too, thinks that natural selection had a lot to do with bringing the primeval soup to life. This outstanding molecular biologist directs Columbia University's Institute of Cancer Research, and the pressing scientific and human problems of cancer allow him only a little time to devote to the origin of life. Nevertheless he has remarkable experiments in progress, conducted by a group of his young colleagues, on evolution before life.

The material they use is a nucleic acid, a portion of the comparatively simple genetic material of a virus known as Q-beta. The team analysed it completely at the start of the research, in order to know the precise sequence of units (A,U,C and G) along the length of its molecule. Like the products of the cloverleaf game in Göttingen, this nucleic acid consists largely of a number of loops. When supplied only with the right chemical building units as 'food' the nucleic acid could in principle reproduce itself, even in a test tube. The experiment is made practicable by the use of an enzyme – a working molecule that helps the nucleic acid to reproduce itself much more rapidly. The enzyme is also part of the nucleic acid's environment, which influences its evolution. But in the soup, evolution of the kind illustrated in the experiment could have proceeded without the benefit of an enzyme.

So the New York experimenters have a nucleic acid proliferating in test tubes. They force the molecule to evolve by poisoning it. An extraneous chemical which attaches itself to one or more points along the length of the nucleic acid impedes the manufacture of new versions of the molecule. One material that will poison the nucleic acid in this way is a dye, ethidium bromide. When added to the mixture of the nucleic acid and its replicating enzyme, it brings reproduction of the nucleic acid almost to a halt. But within a matter of minutes mutant molecules appear that are resistant to the poison and can reproduce merrily. The poison does not itself create the mutant molecules; they exist naturally, in very small numbers at first, made by chance errors during manufacture.

Natural selection in non-living systems is what these experiments demonstrate. Knowing the full chemical structure of the normal molecule, Spiegelman's group is able to pinpoint the changes in the mutants. The poison joins on to the normal molecules at two particular places, and it is precisely there that the mutants have changed. For the normal chemical units at those places, others have been substituted that do not combine chemically with the poison molecule, which therefore fails to attach itself.

In the processes going on at the dawn of life, outright poisoning of a replicating molecule and natural selection of resistant mutants were probably not typical events. But these experiments reveal a quick and subtle response of a mere molecule to an adverse environment. They suggest that, in the soup, molecules able to reproduce could look after themselves and evolve the composition and shape that well assured their stability and reproduction.

The contest of the hypercycles

Besides the nucleic acids, the other essential ingredients of the soup were proteins. To imagine them evolving is more difficult. Unaided, proteins could reproduce themselves only by a very roundabout process. Some proteins have the capacity, as working molecules, to put together a few selected amino acids into short lengths of chain – a start towards making a new protein. Other existing proteins could link the short lengths of chain. But the new proteins built up in this fashion would bear no hereditary relationship to any of the old proteins that made it. A contribution from chance is needed, to set up a cycle in which the newly made protein is for practical purposes the same as one of the proteins contributing to its manufacture, thus creating a reproductive cycle.

This coincidence is highly unlikely. Yet the odds-conscious Manfred Eigen thinks it would occur often enough in aggregations among the vast number of molecules in the soup, if proteins could exploit the information contained in nucleic acids. He visualises chemical cycles going on in a flukey and desultory way, here and there, in the soup. One kind, the simplest, reproduced nucleic acids, which possess information but have very limited chemical function. Another sort of cycle, reproducing proteins, was necessary to achieve chemical function, including the assured reproduction of the information in the nucleic acid.

Picture, then, a rich soup in which untold chemical reactions and independent cycles of reactions were proceeding busily, many of them already influenced by natural selection. Because our very definitions of life are hazy, so is the moment at which life began, at some stage along the chancy path from non-living materials like carbon monoxide gas to plainly living things such as bacteria. But when Eigen addresses himself to the question of what was the minimum assemblage of molecules needed for the leap from chemistry to life, he concludes that it was really rather complex. It depends on a number of these nucleic-acid cycles and proteins coexisting in what he calls a 'hypercycle'.

For a hypercycle to establish itself in a drop of the soup, one of the nucleic-acid cycles with the help of proteins had to make a chemical product – a protein – that would help a second nucleic-acid cycle to work faster. That cycle in turn made a protein assisting a third nucleic-acid cycle – and so on, until the umpteenth protein was assisting the *first* nucleic-acid cycle. All production was from that moment vastly accelerated. Until then, each of the component cycles would have been performing poorly and the proteins would not have been reproducible. Completing the hypercycle greatly enhances their ability both to reproduce and to evolve, by combining the different qualities of proteins and nucleic acids.

The cocktail of necessary molecules was not easy to come by. As there was no previous kinship between the cycles involved, creating a hypercycle by chance required, of course, an astounding coincidence. Yet it falls, according to Eigen, in the category of unlikely events that became inevitable, given enough time. Indeed, he thinks the odds in favour of hypercycles appearing were sufficiently good for them to have been commonplace. Rival hypercycles, using different 'dictionaries' for translating from nucleic acids to proteins, would have competed for the resources of the soup. Their chemical properties ruled out the coexistence of two hypercycles.

The hypercycles had far greater ability to organise the molecular raw materials of the soup, for their own growth and aggrandisement, than anything that went before. In principle, any of them might take over a wide area. But some hypercycles would quickly die out because of internal weaknesses in composition or shortages of key materials. Other hypercycles may have prospered locally for a while, then subsided. Elsewhere, hypercycles may have competed for the same portion of soup. One way and another, natural selection continued to work, favouring the better organised hypercycles until one emerged that took over the world. A chance set of molecules, originally coming together in a single drop of the soup, was, in this view, the grandmother of everything that has ever lived on Earth.

The hypercycle is, of course, nothing but a theory. To some, it may sound too much like an act of creation. But Eigen's is the only theory so far available that really gets to grips with the chemical transition from non-life to life. Because the actual events have left no detailed record in the rocks, we shall probably never know for certain how correct a description it is of what happened when life began. The only test is plausibility, backed up by computer simulations of the processes and perhaps, before long, laboratory experiments with hypercycles.

Basic characters

One of the advantages possessed by the 'winning' hypercycle was probably an ability to wrap itself up in small packets – the first living cells. It may have been done by forming oily blobs in the water, or by making fatty materials that formed a skin or membrane encasing the hypercycle's cocktail of essential chemicals. In some such way the hyper-

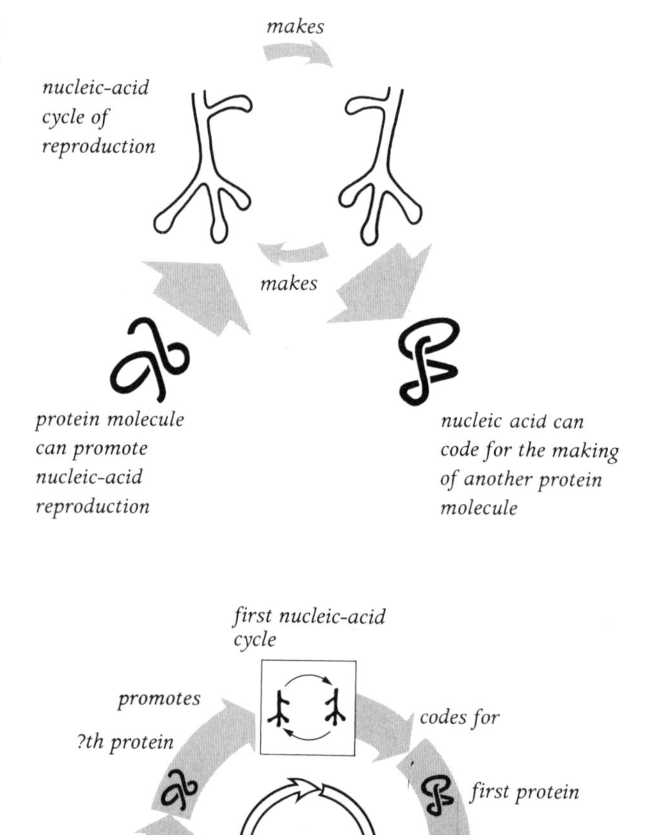

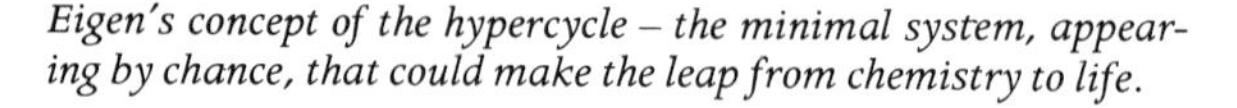

Eigen's concept of the hypercycle – the minimal system, appearing by chance, that could make the leap from chemistry to life.

cycle escaped poisoning by extraneous chemicals and avoided scattering its key products. As each cell grew, feeding on the soup, it would break up when it reached a certain size, forming new cells. There would be strong natural selection in favour of those cells whose 'daughter' cells managed to retain all essential molecules.

The origin of the first individuals, microscopic cells, created ideal conditions for natural selection to favour minor or major improvements both in internal molecular arrangements and in the organisms' ways of making use of the soup. Internally, the next big step was towards more rationalised interactions between nucleic acids and proteins. Part of this was the adoption of a common code for translating information in the nucleic acids into protein working molecules. Without it, the component cycles of the hypercycles could only confuse one another; with it, the streamlined methods of making protein, used even by the simplest organisms of today, could originate.

Exactly the same genetic code serves every living thing, from bacteria to whales. That means we are all descended from a single very early population of living cells in which this code established itself. Another implication is that those organisms had already settled on the favoured twenty amino acids that would always be used for making working molecules. Any later departure from the code would produce chaos in a cell. For example, the sequence of nucleic-acid units GCA codes for the amino acid alanine. If it were read instead as 'glycine', then every working molecule in the cell would suffer multiple mutations, with glycine replacing alanine. The effects would be fatal.

Until about this stage, 'nucleic acid' had meant primarily the molecules of RNA (ribonucleic acid). These are still used as agents of the genes in living cells, and as genes by some viruses, such as Spiegelman's Q-beta. But the genes of organisms are now all embodied in DNA (deoxy-ribonucleic acid). At the origin of life RNA had advantages, being easy to make and also adopting those variable cloverleaf shapes which gave natural selection something by which to distinguish 'good' and 'bad' molecules. To the outside world, the elegant double helix formed by matching strands of DNA always looks the same. So natural selection among molecules could hardly operate on DNA. Only when the first rudimentary organisms existed would the advantages of enshrining the genetic messages in DNA become apparent in the cells' performance.

Bacteria feeding on the chemicals of the soup were almost certainly the first organisms. Like bacteria today, they would take existing energy-rich materials, ferment them to extract the energy by promoting particular chemical reactions, and use pieces of their food to build their own cells. Between the various kinds of bacteria, then as now, an immense variety of chemical talents would have been represented, with working molecules adapted to making the best use of particular ingredients of the soup. One example of microbial specialisation has aroused the interest of investigators of life's origins. It concerns organisms living at the base of the walls of Harlech Castle, in Wales, which thrive on ammonia accumulated from hundreds of years of urination. They resemble nothing more closely than microbial fossils found in Canadian rocks 1700 million years old.

Much earlier even than that, the very success of the bacteria created the first major crisis of life on Earth. To begin with, there had been an immense stockpile of chemical resources in the soup – food for the hypercycles and the first organisms. But once active biological chemistry began in earnest these re-

Fossil bacteria about 2000 million years old, revealed in Canadian rock by an electron-microscope photograph.

sources were far from unlimited. The bacteria could not survive indefinitely if the only source of energy was the diminishing remains of the soup. The earliest life might easily have eaten itself out of existence.

A change of menu

With the soup becoming thinner, natural selection strongly favoured any organisms that could do without it. The only other convenient source of energy was sunlight. Some of the early microbes accumulated within their wrappings very elaborate machinery for absorbing the energy of the Sun's rays in a pigment and using it to make their own chemicals as food. These first plant-like organisms could accomplish, in a systematic fashion, what had been done randomly and inefficiently by the 1000 million years of lifeless chemistry that made the soup. The origin of photosynthesis, in which sunlight fosters a reaction between carbon dioxide and water, to form sugars and oxygen, marks the start of self-sufficient life on Earth. At this stage, fortunately, speculation gives way to tangible evidence – the remains of real organisms preserved in ancient rocks.

The oldest certain fossil remains of living things occur in rocks in the region of Barberton in the eastern Transvaal. The rocks belong to what the geologists call the Fig Tree series and they are 3200 million years old. They contain nothing more than microbes: bacteria looking very like some modern bacteria and microscopic plants indistinguishable from the blue-green algae, the most primitive of all plants that still prosper today. Chemists and biologists who had been speculating about the origin of life could be happy when the Fig Tree fossils turned up in the 1960s. The mixture of bacteria and primitive plants fitted well with their ideas of what life must have been like at the end of the Age of the Soup.

These organisms mark an early phase in the Age of Microbes. Thereafter, and for more than half of the Earth's long history, from over 3200 to less than 800 million years ago, the microbes had our planet to themselves. It is true that the blue-green algae began to grow in strings. And larger structures appeared: rock-like mounds called stromatolites, which blue-green algae still build today, occur in very old rocks. But you need a microscope to make out the individual organisms responsible for their formation. At first sight, the Age of Microbes was a great longueur with life amounting to little more than a green tinge in the water.

As the microseconds of chemical reactions ticked away, though, and as seconds became hours, hours turned to years, years became centuries, and millions of centuries slipped past, the microbes were steadily accumulating a vast repertoire of chemical and organisational skills. On these skills depended the evolutionary burst into the Age of Visible Animals, 600 million years ago. On the tricks devised in microbes in the remote past we still depend utterly in our daily lives. With every breath we take, in fact: the exploitation of oxygen began in microbial defences against oxygen.

Early life had escaped from the dangers of the exhaustible soup only to run into a second major crisis. The plants threatened to poison themselves, and the bacteria as well, with the oxygen that they made as a by-product of their growth. Oxygen had not existed in any noticeable quantity since the world began; had it done so, it would have prevented the origin of life. Living material is, by chemical nature, stuff in which oxygen is scarce; oxygen tends to destroy it, in the metabolic equivalent of a bonfire.

Locally at first, in the immediate presence of growing blue-green algae, free oxygen created severe

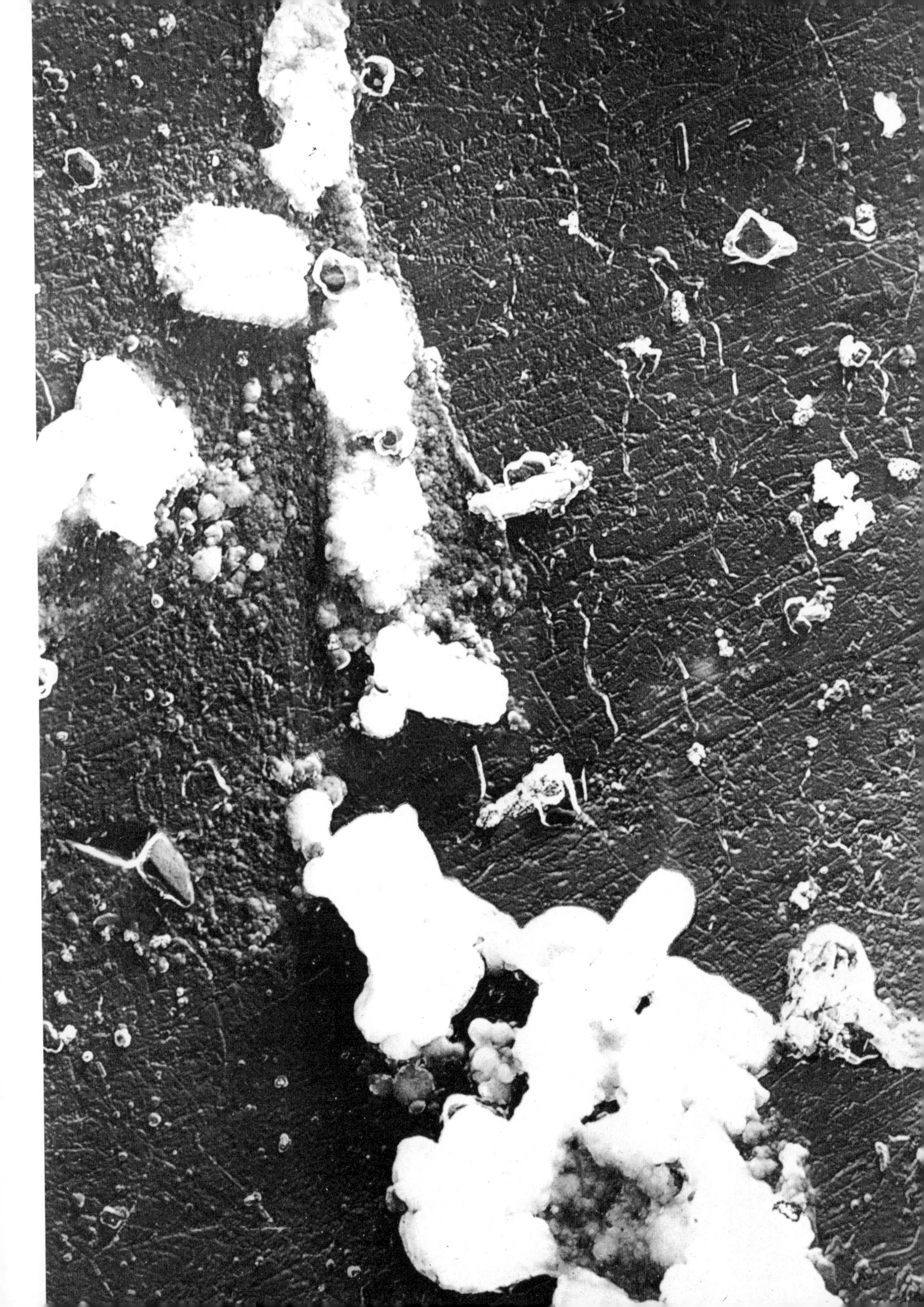

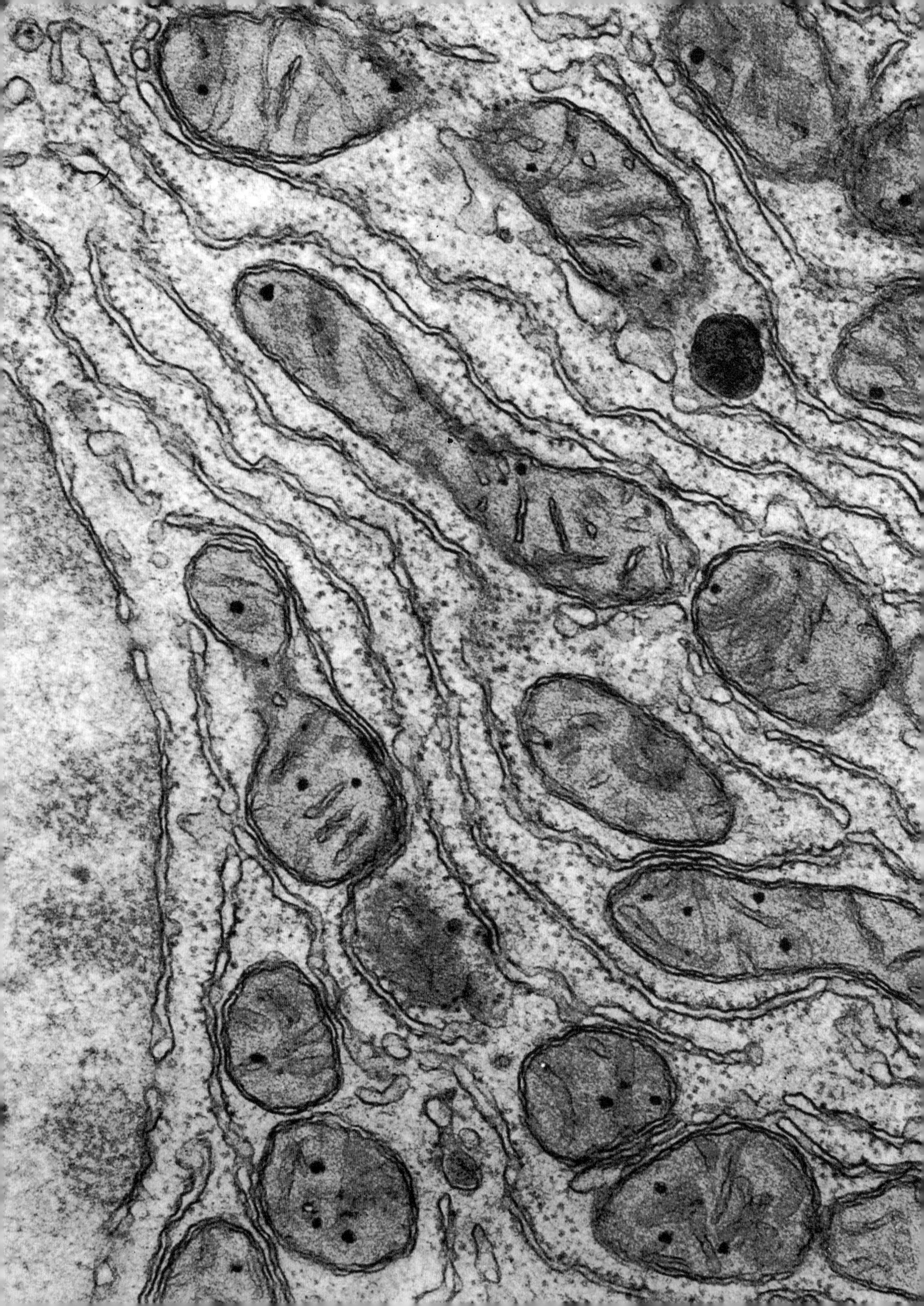

In part of a liver cell of a rabbit, the electron microscope reveals the bodies – the mitochondria – that handle oxygen.

problems for the primitive organisms. Many, indeed, must have been killed. Strong natural selection favoured bacteria possessing any chemical means of coping with oxygen. Some never acquired such defences; their descendants today constitute a class of bacteria which shun the air. Others achieved the ability, now shared by all plants and animals, not merely to resist the harmful effects of oxygen but to turn it to advantage as a source of energy. That was not as easy as it sounds. To capture in a useful fashion the energy released when oxygen combined with food, elaborate combinations of working molecules were needed. Cytochrome c, which figured prominently in the previous chapter, was only one of the complex molecular machines invented at the time.

Modernising the cells

In the oxygen crisis, evolution apparently took a remarkable short-cut. One primitive cell could obtain the oxygen-using machinery that it urgently needed by swallowing whole another microbe that already possessed the machinery, and then keeping it alive as a permanent lodger. That, at any rate, is a theory now rapidly gaining support. It says that little bodies within our own living cells, the so-called mitochondria which exploit oxygen for us, are direct descendants of bacteria that took up residence in other cells during the Age of Microbes.

Biologists have wondered about this curious possibility since they first spotted mitochondria in living cells in the nineteenth century. They saw their resemblances to bacteria both in appearance and in their habit of dividing and reproducing, independently of the life cycle of the cells to which they belonged. Weighty new evidence for bacterial origins, and a prime contribution from molecular biology to

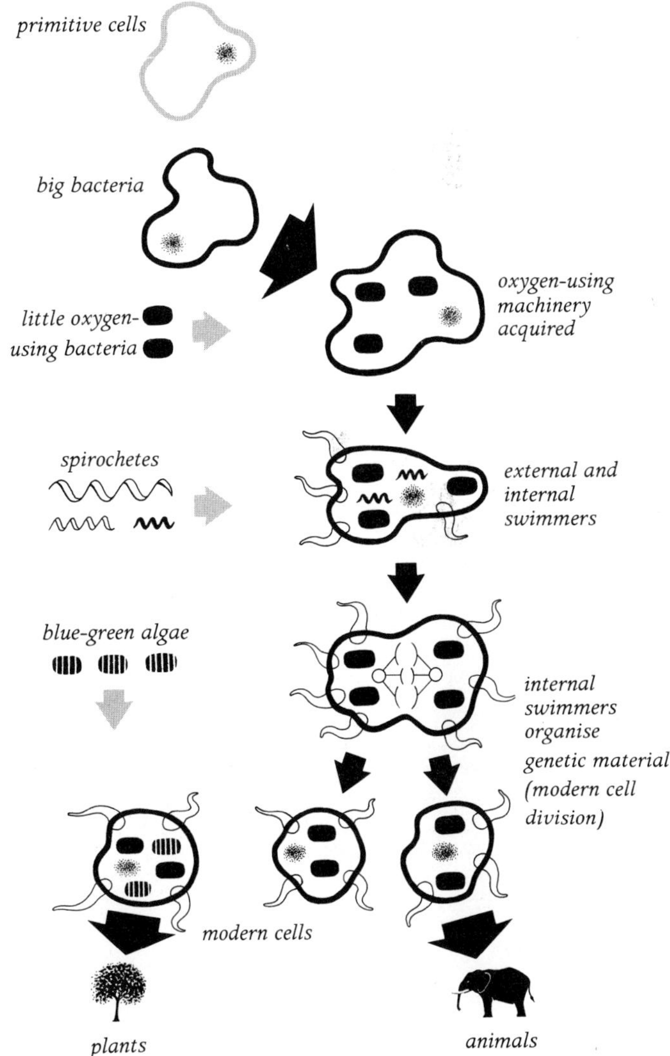

How primitive microbes may have clubbed together to form more modern cells capable of evolving into plants and animals (after L. Margulis).

A modern cell possesses machinery (right) that takes charge of the chromosomes during cell division and the production of sex cells. This machinery may be derived from the most primitive kinds of mobile microbe, like the spirochete (below). The bottom photograph shows spirochetes that have attached themselves to the surface of a microscopic animal, pushing it around like little oarsmen.

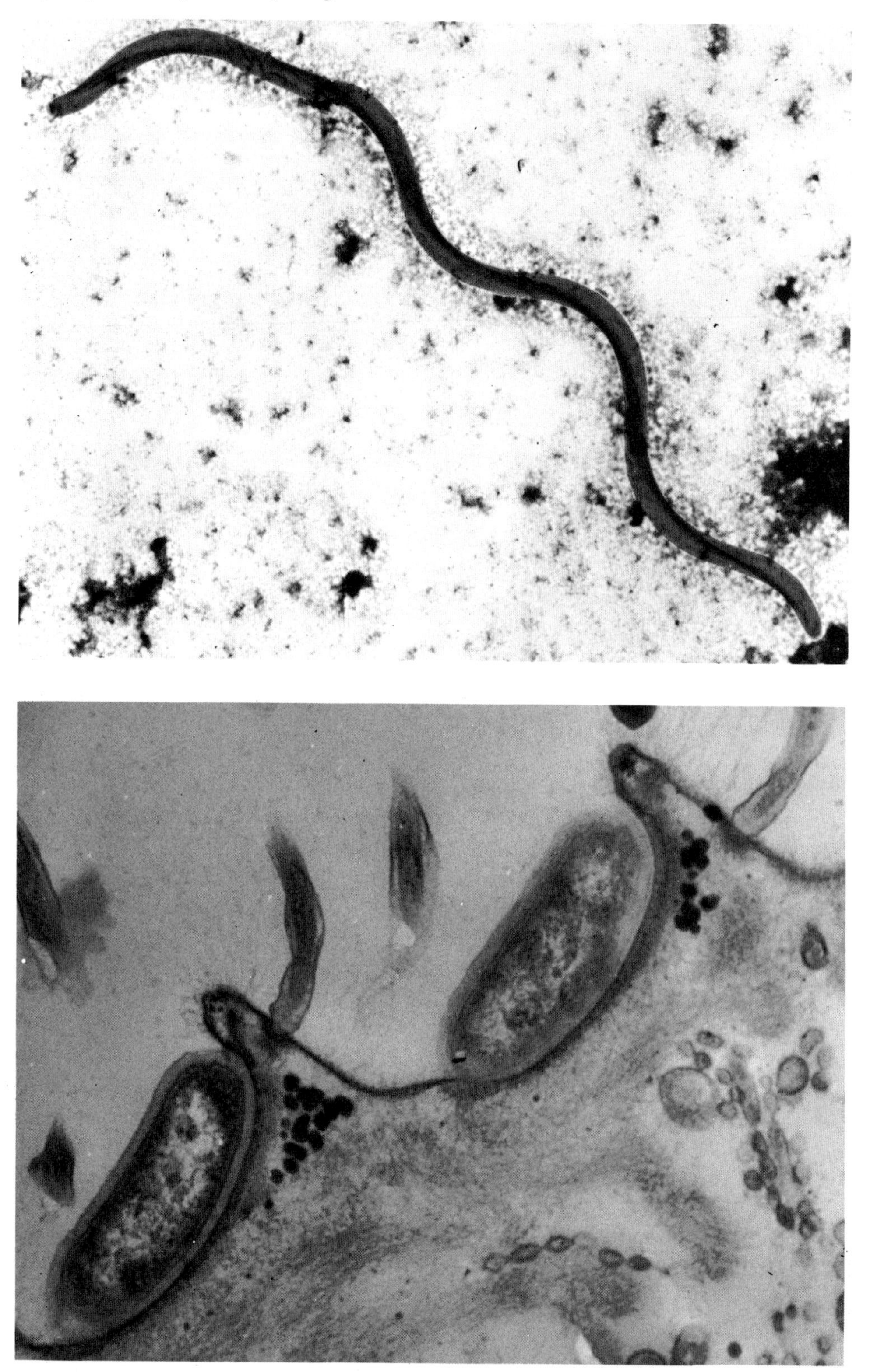

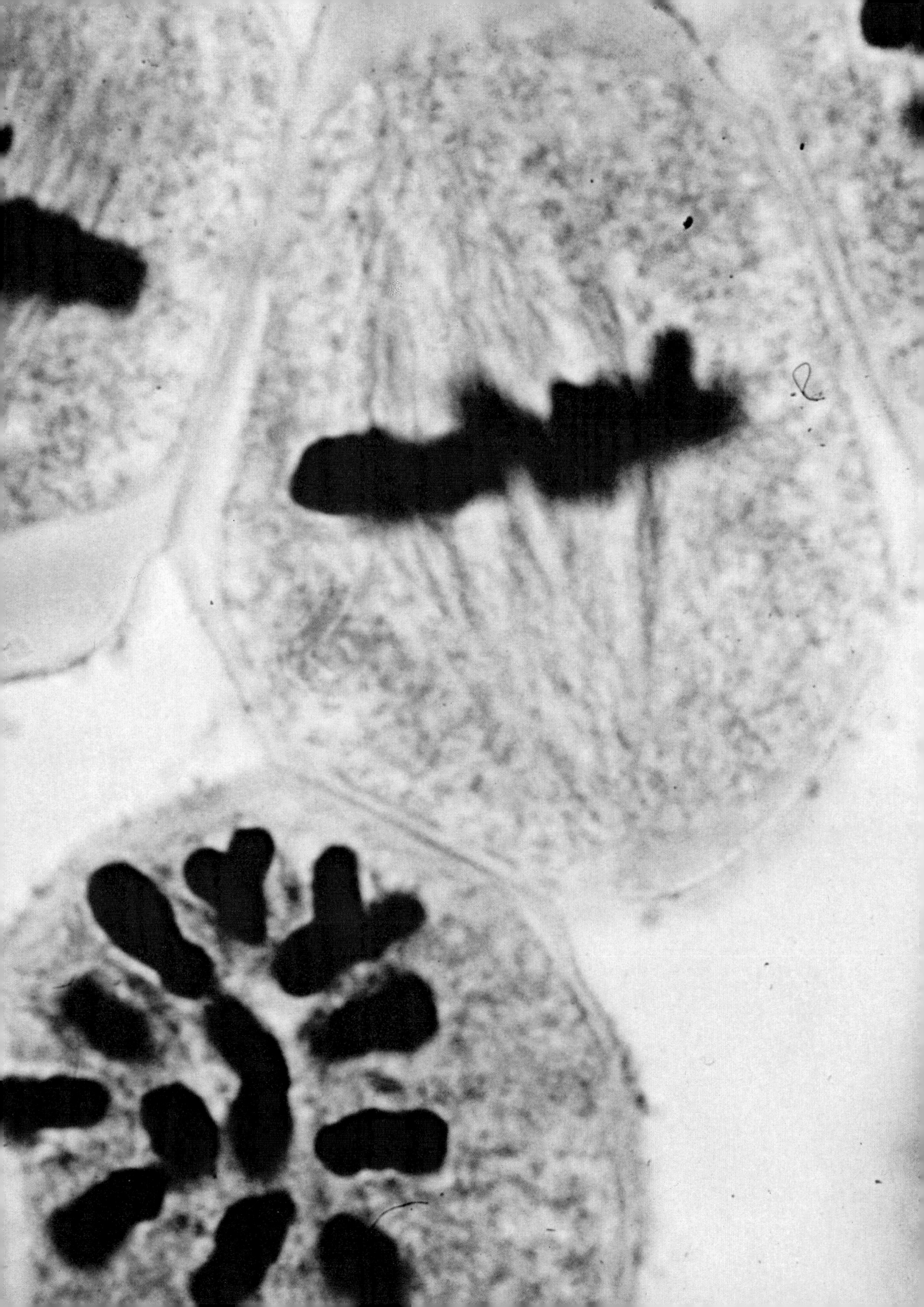

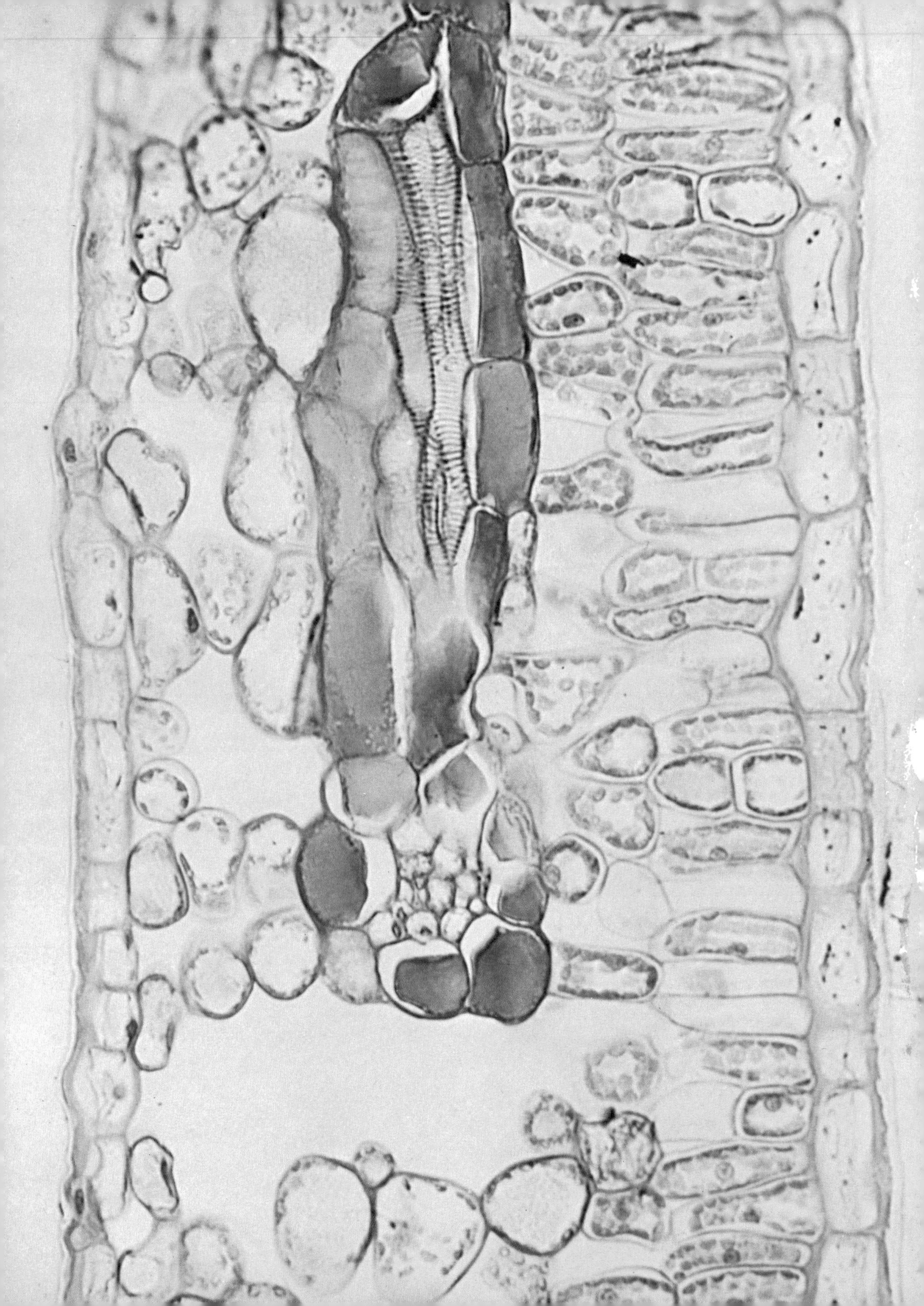

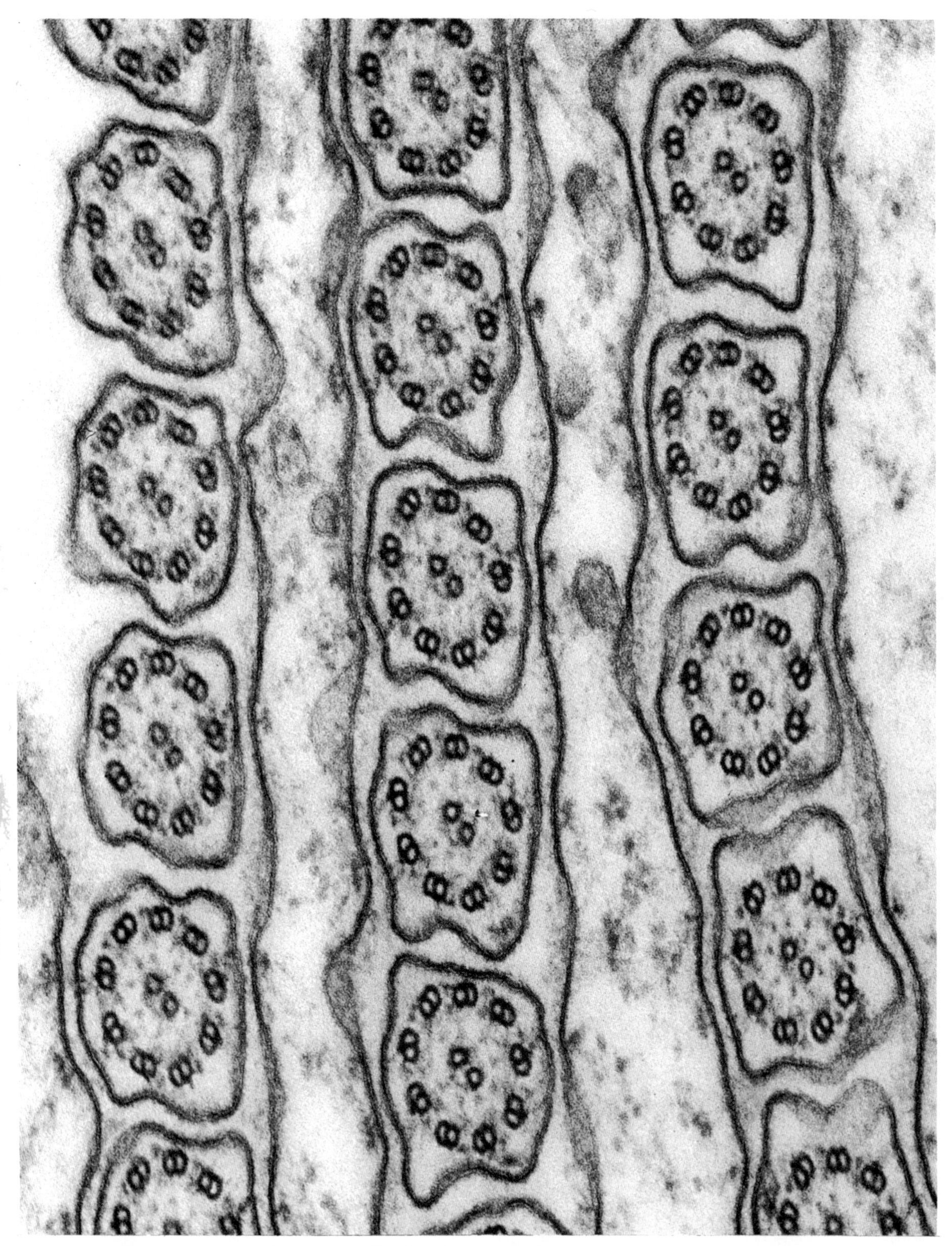

The bodies (left) that capture sunlight in the cells of modern plants are normally green, but here have been stained for examination. They may be remote descendants of blue-green algae that long ago took up permanent residence inside other microbes. The cross-sections (above) of some of the flagella that confer powers of movement on cells show a characteristic ninefold ring of tubules. Similar ninefold structures are found in the internal machinery of many different cells.

evolutionary science, comes with the discovery that the mitochondria still carry genes of their own, very like the genes of bacteria. Yet the mitochondria live outside the nucleus of the cell where most of the genes reside.

Within the cells of your body the mitochondria carry on their own sex life. They mate and exchange genes, operating a system of heredity and evolution quite distinct from the 'orthodox' genetics of the cell nucleus. What seems to have started as an optional life for bacteria in other cells has now become obligatory, because the genes are incomplete and the reproduction of mitochondria depends upon assistance from the genes of the nucleus. Again, like bacteria, mitochondria are vulnerable to antibiotics – and researchers are concerned about possible harm to these essential machines of human cells by present carefree uses of antibiotics.

Primitive cells clubbing together, to make cells of more modern types, may have achieved several of the most basic steps in evolution. That the oxygen-handling bodies within the cell, the mitochondria, are descended from bacterial lodgers of the distant past is only one aspect of a broader theory, for which Lynn Margulis of Boston University is the most outspoken advocate. I found her managing to combine the raising of four children with her pursuit of a controversial line of research. Her greatest problem has been what she politely calls a lack of communication' among specialists in different branches of the life sciences. She has incurred the displeasure of some orthodox biologists, but others are beginning to look with favour on the theory.

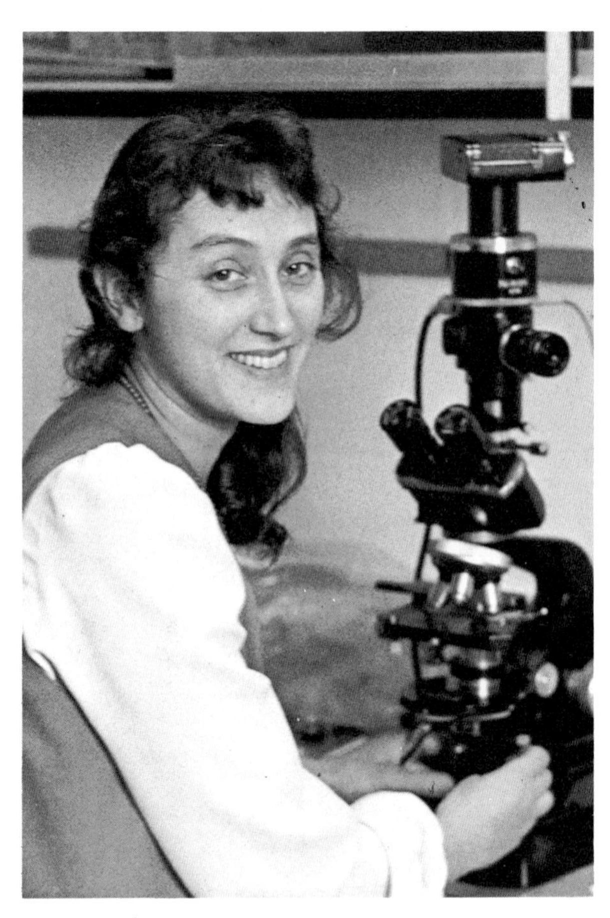

Lynn Margulis.

Cells of the modern sorts differ radically from the primitive cells of bacteria. For example, many possess independent powers of movement conferred on them by little whip-like appendages, known as flagella or cilia. But the chief features distinguishing modern from primitive cells are the way the genetic material is organised, in a package in the nucleus of the cell, and the machinery that carefully manipulates the chromosomes carrying the genes when the cell divides. Margulis says that cells have acquired these and other characters from alien cells taking up residence inside them, to the mutual benefit of the incomer and the host. She envisages the following events.

In the oxygen crisis created by the blue-green algae, some bacteria which did not possess the oxygen-using machinery took in a number of partners that did. The resulting cells, like animals today, depended on external food supplies. Their mobility in searching for food was improved by taking on outboard motors. The motors were microbes known as 'spirochetes'. These wriggle about their business and today include the organism that causes syphilis. The little swimmers attached themselves to the host cell and pushed it along. Such was the origin of the flagella and cilia. The theory also supposes that some of these swimmers went to live right inside other cells. If so, they now contribute to the workings of the modern cell; in particular, they marshal the chromosomes and manipulate them during cell division and in the making of eggs and sperm.

Sex becomes interesting

To the resulting 'supercells' Lynn Margulis accords a central place in evolutionary history. From them three great kingdoms of life are descended. First, the animals have evolved from cells possessing, as just described, the oxygen-using bacteria and the internal and external swimmers. Then the fungi, such as mushrooms and yeast, retain the function of the swimmers internally, for cell division, but have lost (if they ever had it) the power of active movement. The third great kingdom, the higher plants, originated from cells that took in yet more lodgers. Small blue-green algae, assimilated into the supercell, have conferred on it the ability to make its own food supplies from the energy of sunlight. The machinery of these algal guests, now streamlined, is visible today under the microscope as small green bodies, the chloroplasts, living inside plant cells. They make the living landscape green, and are the ultimate source of all our food.

The possibilities of permanent lodgers are manifest in organisms of our time. Some small animals have their own internal gardens of algae that grow by sunlight and supply the animals' needs for food while the algae enjoy shelter. Margulis cites particularly a curious microbe that lives, of all places, in the digestive tract of an Australian termite. It is driven around by large numbers of the primitive swimming spirochetes. Established in its membrane, they beat their tails as rhythmically as galley slaves, so that their hosts travel along straight tracks.

Whether or not the course of events was exactly as Margulis says, the modernisation of cells remains one of the major steps in the history of life on Earth. Recent claims of evidence for organised cell nuclei of the modern type in microbial fossils of 1000 million years ago are now doubted. The features seen could just as well be other parts of the cells that congealed after death. Even so, that date may be about right, judging from molecular comparisons of cytochrome c, as described in Chapter 3. If the average rate of molecular evolution in the Age of Microbes was the same as it has been since then, the most recent common ancestor of animals, plants and fungi lived about 1200 million years ago.

The modernisation of sex was an important consequence of these advances. At its simplest, sex is just a

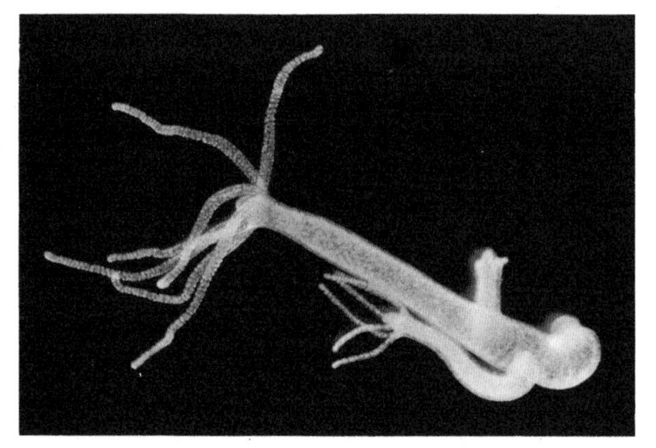

The small water-animals, hydra, have astonishing powers of regeneration. The photograph (left) shows a normal hydra, with two buds that will form new animals. If hydra are broken up into individual living cells (first photograph, right) and those are then amassed into a sausage, the cells reorganise themselves into new hydra. The two lower photographs show this process starting, three and four days after mincing. After several weeks, normal hydra bud off from the aggregate.

means of transferring spare copies of genes from one individual to another. Among bacteria two cells can come together, a tube forms between them, and the 'male' injects some genes into the 'female'. In such primitive organisms sex may have originated as a means of making good any genes damaged by mutation; sex is not needed for the reproduction of bacteria, which simply grow and divide. But nowadays bacteria that have survived the onslaught of an antibiotic, and possess genes conferring resistance to the drug, can by sex transfer those genes to other bacteria.

Among cells with the more modern machinery for organising the genetic material, reproductive life could become more interesting. In the ordinary division of such cells, the genes are first duplicated. Then the machinery (that originated, according to Margulis, from swimming lodgers) apportions a complete set of genes to each daughter cell. The crucial variation, for modernised sex, was to divide cells in such a way as to give half the set of genes to each resulting cell. When two complementary sex cells came together, one from 'mother' and one from 'father', they recreated between them a complete set of genes to form a new organism. Sex and the endless shuffling of genes from two parents accelerated evolution by creating a greater variability between individuals on which natural selection could work. Parent organisms also had to be in the right place together, and show the right patterns of behaviour, if successful reproduction were to occur by mating. So sex was a powerful factor favouring evolution towards more discriminating and complex behaviour.

But we are still in the Age of Microbes. Yet another great innovation had to occur before the Age of Visible Animals could commence. Masses of cells had to be able to live together, not just as colonies of identical cells but as complex organisms in which different cells had different tasks, as in a modern animal. How was this trick accomplished? That question leads us into what is perhaps the most exciting area of biological research of the early 1970s. The subject under investigation, the growth and development of a complex organism, is fascinating in its own right. It may also help the theory of evolution over some awkward hurdles.

Why mince hydra?

Anyone insensitive to the wonders of life should be required to consider what was involved in creating him from a newly fertilised human egg, a tenth of a millimetre in diameter. The mere increase in bulk is the least astonishing aspect of it; bacteria can multiply themselves a good deal faster. What distinguishes us from bacteria is the exquisite organisation imposed upon the mass of proliferating cells. Spontaneously each part of the body adopts the form and character appropriate to its function, as cells take on specialised tasks in muscle, nerve, skin or blood. Out of the formless blob of the egg come elaborate yet compact organs: kidneys, for instance, and the eyes which are more like computers than cameras. Wiring and conduits more ramified than the communications and transport systems of an entire nation are packed into a volume smaller than a cupboard.

The discoveries in molecular biology and the 'reduction' of heredity to strings of chemical units heighten rather than diminish the glory of it all. So long as genes and protoplasm were unanalysed any kind of quasi-magical property could be ascribed to them. But now we know what the genes are made of we can see that, superimposed on their essential simplicity, is an elusive system of control. Like the genes themselves, it must rely upon the laws of chemis-

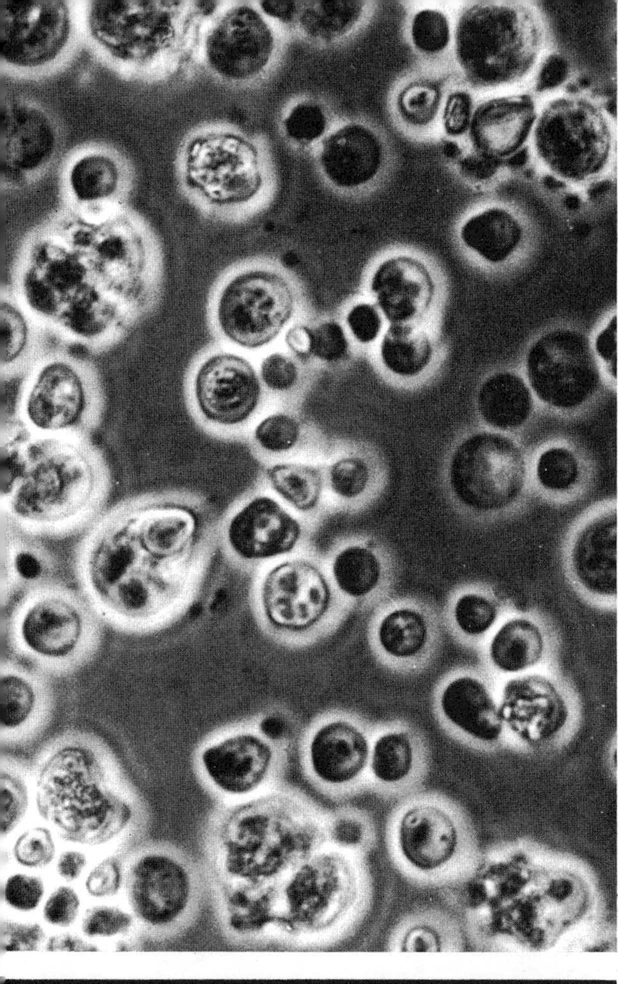
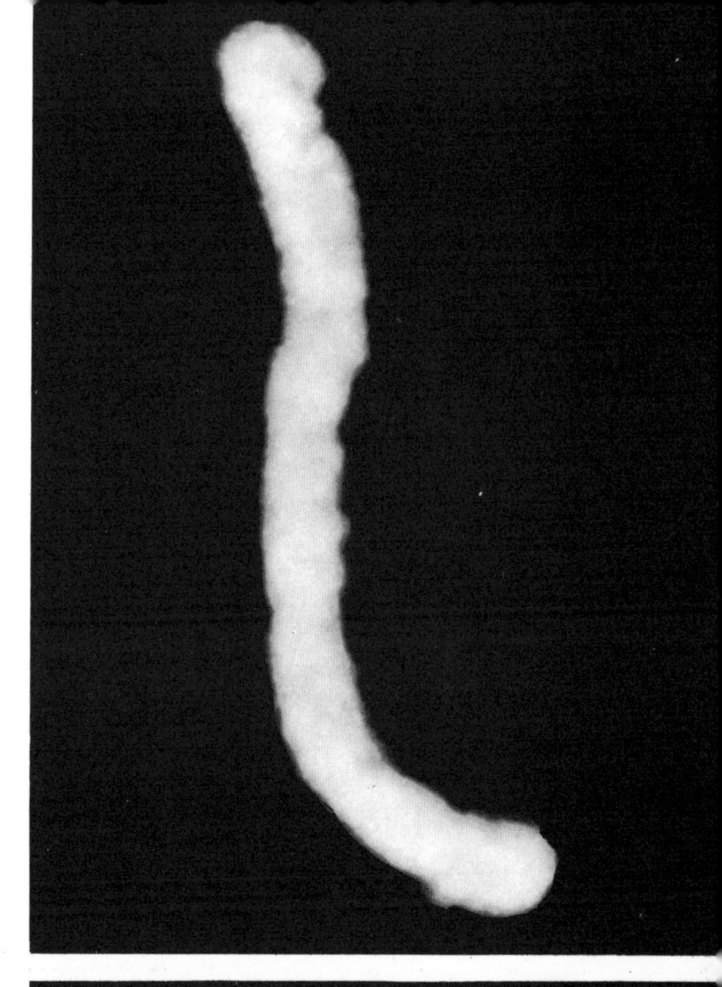
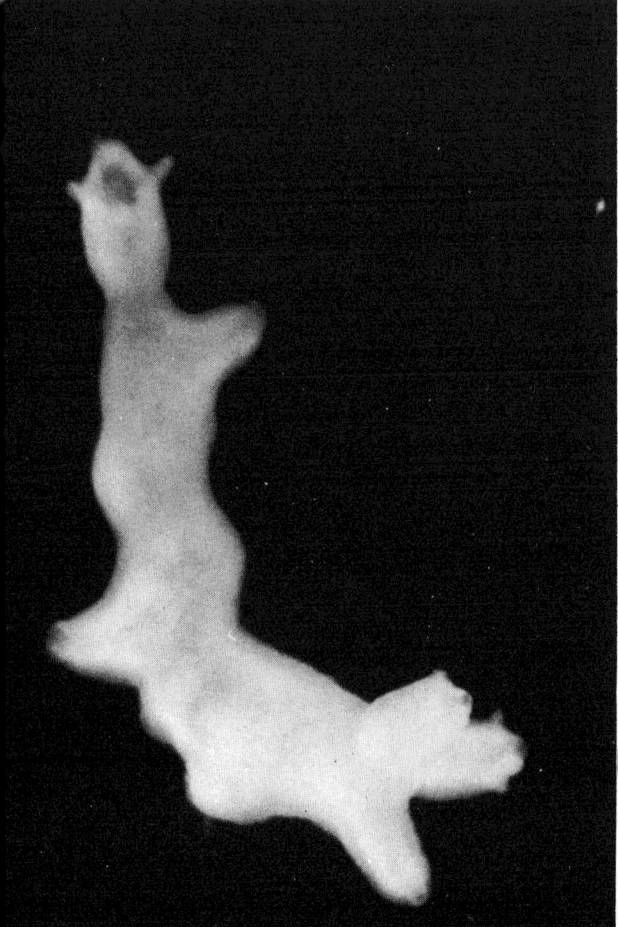
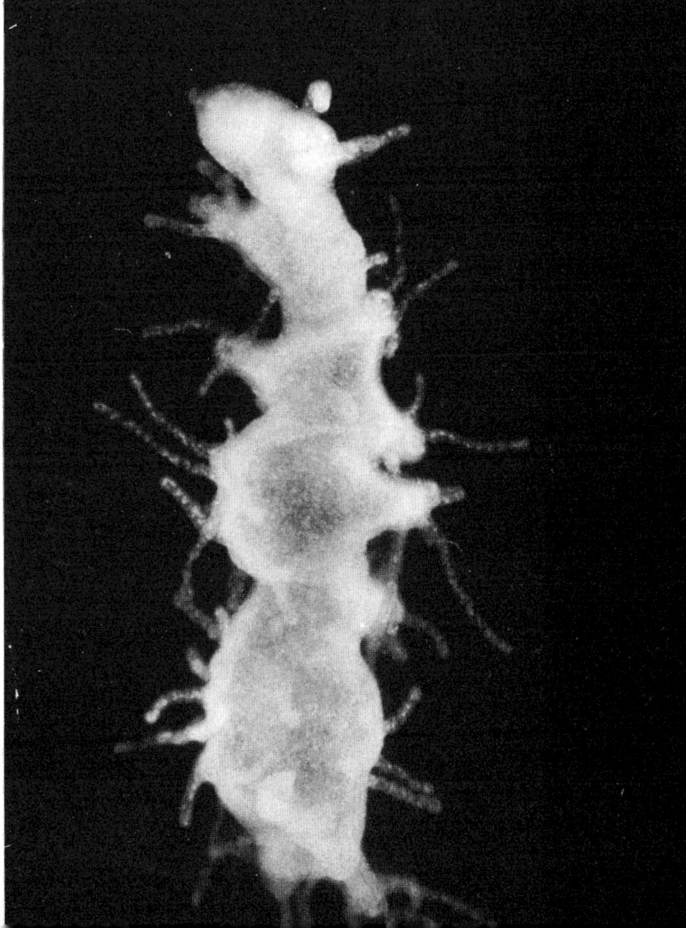

try. What distinguishes muscle from brain is that, within the cells that compose them, different sets of genes are active while others are suppressed. Almost every cell of the body contains the complete set of genes of the egg from which it is descended. Normal frogs have been grown from the genes of cells from the frog's intestine. And in teratoma, a form of cancer, cells of the wrong sort – of teeth, for example – grow in the affected part of the body.

The controls that make such monstrous confusions the exception rather than the rule have themselves, of course, evolved over many millions of years. The opinion grows stronger that changes affecting not the ordinary 'producer' genes but other genes that control them hold the key to some of the more dramatic appearances of new players in the life game. Some biologists now suppose that, even in animals of great complexity like ourselves, the techniques of control remain essentially simple and akin to those used in the simplest animals.

Alfred Gierer.

The hydra is a little animal living in water. Although only a couple of millimetres long, it shares a proud name with a monster of Greek mythology with the alleged ability to grow new heads in profusion whenever an old head was destroyed. The little animal of real life also has remarkable powers of regeneration, which has made it ready prey for biologists wanting to investigate the control of growth and development. The severest treatment is that meted out to hydra by Alfred Gierer and his colleagues of the Max-Planck Institute for Virus Research at Tübingen in Germany.

The biologists break hydra into their constituent cells, by repeatedly blowing them through tubes which are too narrow for them. Not many animals could survive such treatment but the hydra does. If the dismembered cells are brought together in a loose mass, they assemble themselves into a new animal. Although rather misshapen, within a week it is capable of feeding, and before long is producing perfectly normal offspring. In this ability the hydra surpasses anything envisaged for its mythological counterpart. On the other hand, it is more fussy about how many heads it has. And therein lies the point of the experiments.

Far from being a circus trick for aquatic animals, these German experiments have been helping, since 1972, to bring out fundamental rules about the way in which the cells of minced hydra organise themselves. Regeneration is possible with suitable numbers and types of cells derived from various portions of the animal's body. Whatever portions are used, the cells that were nearest to the head of the original owner are always the ones that give rise to the heads of the regenerate animal.

Instructions for growth

This work with minced hydra tallies with somewhat less drastic manipulations of the same animals at the Middlesex Hospital Medical School in London. Lewis Wolpert, who leads the work there, was formerly an engineer, which colours his views of biology. Living cells are not intelligent beings capable of reading an elaborate blueprint to find out what function they are supposed to perform within the overall design of the body. At best they are little robots, programmed to make a limited range of chemical decisions in response to information about where they are. Experimental work by Wolpert's colleagues at the Middlesex, first with hydra and most recently with chicks, sustains his conviction that the mechanisms cannot be very complicated.

New heads were grafted on to hydra, and the old heads were cut off at a later stage. The old head

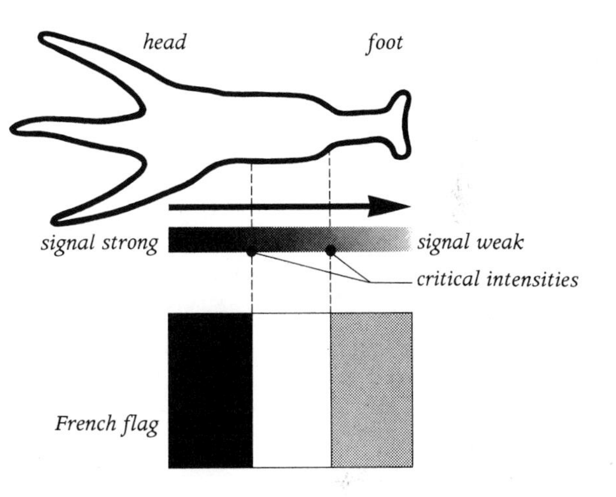

The French flag problem. A cell deciding to which part of the hydra's body it belongs is like a person in a crowd which is trying to make a French flag by holding up coloured papers. To know what colour to hold up, the person must find out where he is in the crowd. The hydra evidently solves the problem by a chemical signal that gradually diminishes in intensity from head to foot.

Research into the mechanism by which the growth of a chick's wing is controlled. At the Middlesex Hospital Medical School (right) this work includes transplanting the bud out of which the wing grows, from a younger to an older embryonic chick.

Lewis Wolpert.

could prevent the new head from 'taking' if it were left in place too long or if the new head were grafted too close to it. A chemical influence evidently emanates from the hydra's head. Any cell can tell its situation (near the head, near the tail or somewhere in the middle) according to the strength or weakness of this chemical.

Wolpert compares the cell which has to decide what kind of cell to be to a person in a crowd who is supposed to be helping to represent a flag. He is supplied with pieces of paper of different colours but, to decide which colour he is meant to hold up, he must know where he is in the crowd. A simple system, such as numbering from one side of the crowd, would suffice for making something as straightforward as the French flag (blue, white and red stripes). A low number means 'hold up the blue paper'; high numbers, 'red'; intermediate numbers, 'white'. The simple hydra is not so very different from a French flag.

Higher animals are patterned in complex ways, more like the Stars and Stripes or the Union Jack. Their cells need information about their positions reckoned in more than one direction. Wolpert's experimenters turned to the wings of the growing chick. The cells of the wing may obtain part of their information in something like the hydra fashion, from a source lying at the rear of the bud on the side of the developing body, from which the wing grows. From this, each cell evidently learns whether it belongs on the leading edge of the wing, towards the back, or in the middle. At any rate, if the rear part of a wing bud is grafted in front of another chick's wing bud, the latter chick will grow a double-banked set of wing bones.

For instructing its cells about their positions along the length of the wing the chick embryo seems to use a quite different system. According to

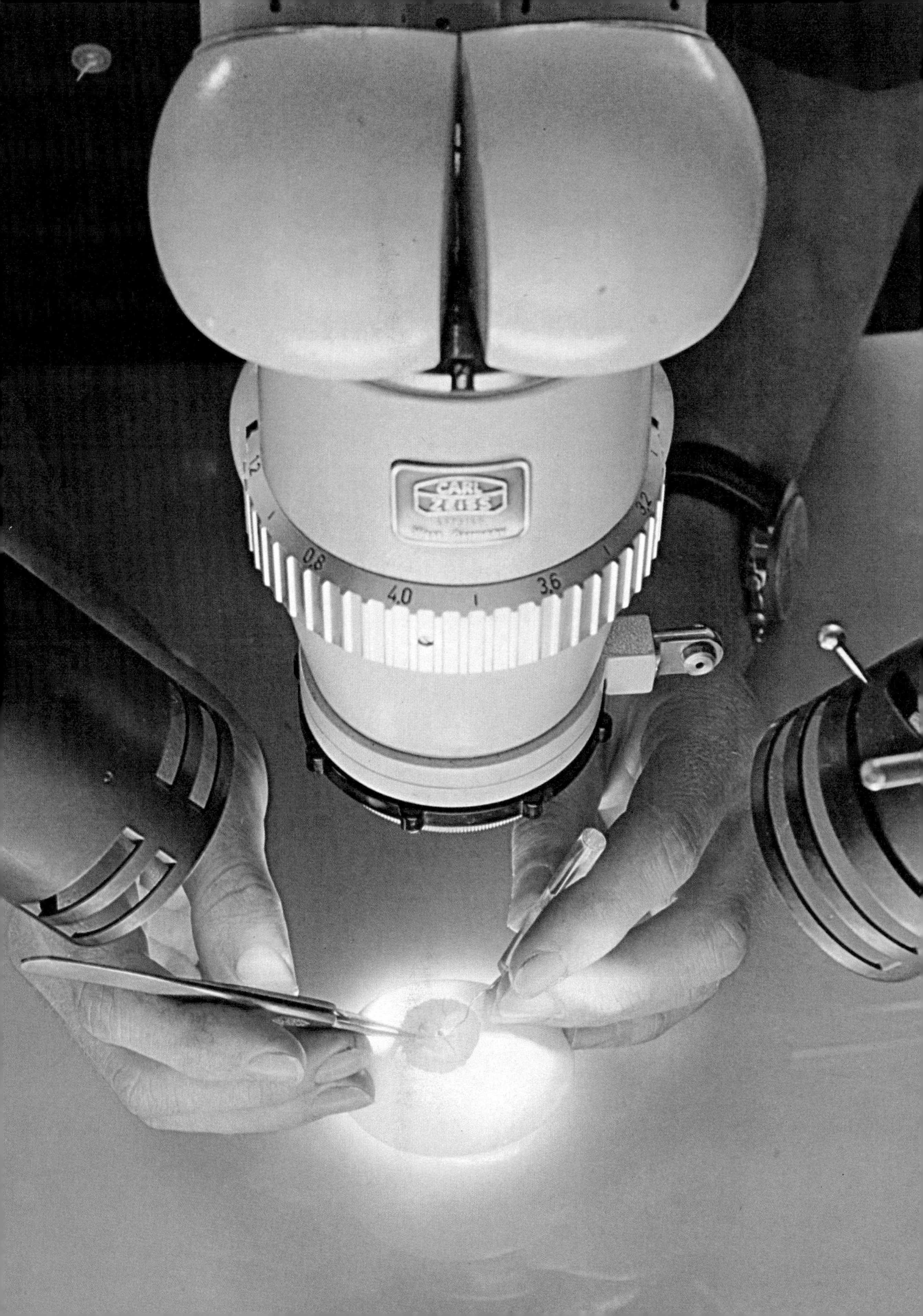
CARL
ZEISS
0.8
4.0
3.6
3.2

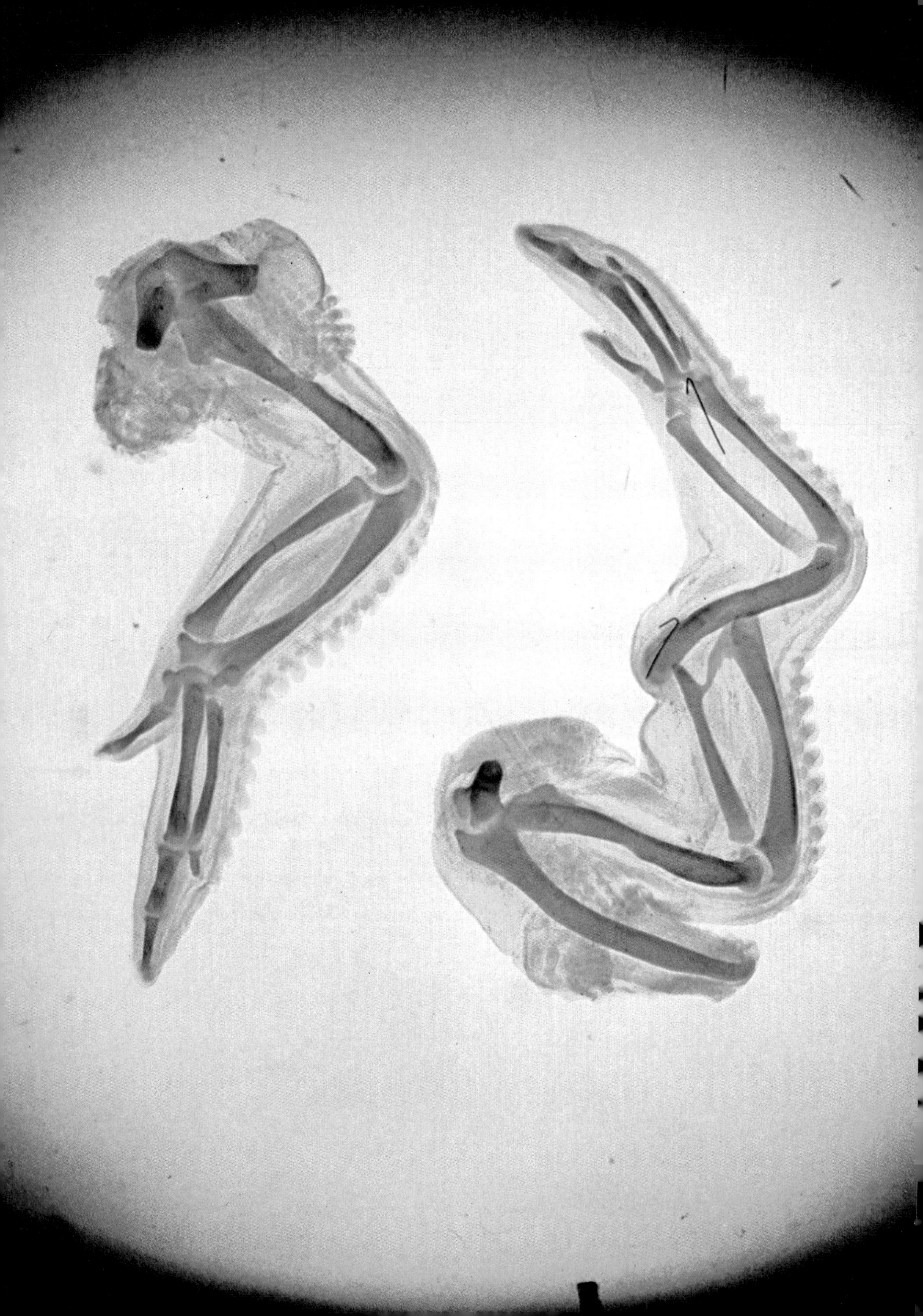

The wing resulting when a wing-bud is transplanted from a younger to an older chick is compared with a normal wing (left). A renewal of growth results in a repetition of the upper arm and forearm (lower diagram below). Another experiment (upper diagram) suggests that the control of growth across the wing may be governed by a mechanism similar to the hydra's. Experiments like this, carried out in animals, are necessary for trying to understand grievous errors that sometimes occur in the growth of human babies – in the Thalidomide tragedy, for example.

a theory broached by Wolpert in 1973, each cell discovers its position in this other direction, from the time that elapses before they emerge from the wing bud. More precisely, Wolpert calls a region in the bud 'the progress zone', and the time spent in this zone is what matters. As the wing grows the bud moves outwards on the end of it. The cells that spend least time in it give rise to the parts of the wing nearest the shoulder; the cells of the wing-tip remain the longest in the progress zone and emerge last. How do cells tell the time? An obvious clock is provided by the repeated division and redivision of cells in the growing embryo. The last cells to emerge from the progress zone have divided most often during their period in it.

One prediction from this theory is that, if a new bud on a young embryo is transplanted to the end of the growing wing of an older embryo, the latter chick will grow a wing doubled lengthwise. The new bud will repeat the bones already made by the original bud. This experiment has been done and the result is exactly as expected; it is illustrated in the diagram and in the colour photograph opposite.

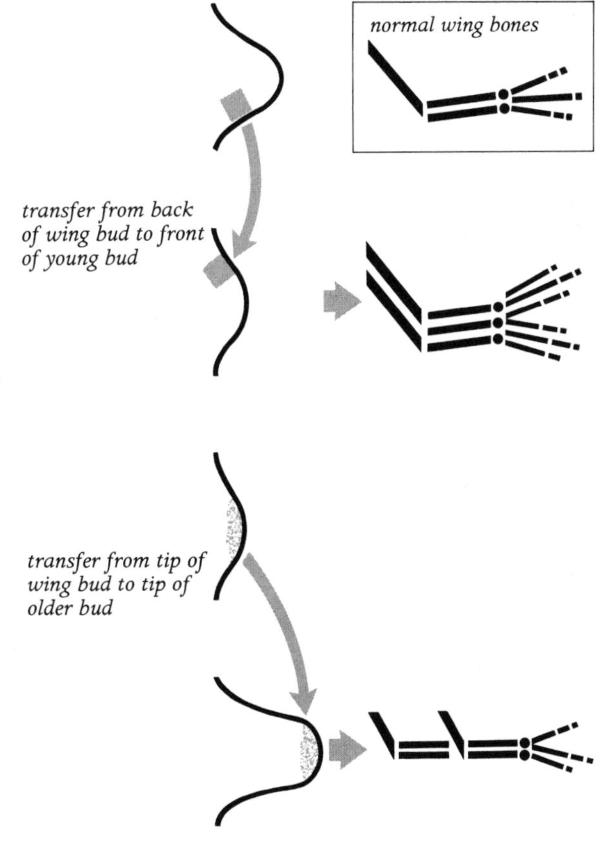

The genetic computer

Other research teams pursue the systems that control the growth of a complicated animal down into the genes themselves. Relevant here is the extraordinary exercise in computerised anatomy which, as mentioned in Chapter 1, Sydney Brenner at Cambridge is using, to trace the effects of mutations on growth in the nervous system of a nematode worm. But the government of the genes, by the genes, for the genes, is widely regarded as the prime interest in current molecular biology. Depending on whereabouts in the body a cell lives, it must be able to switch on relevant genes and permanently inactivate all the inappropriate ones. On a continuing basis, the cells must be able to regulate their supplies of working molecules, so that if they have enough of a particular material for the time being, the genes coding for its manufacture can be switched off. The fact that cancer involves a breakdown of the cells' genetic discipline gives a broader significance to this fundamental research.

The cells of higher animals possess a great deal of genetic material (DNA) which is not directly involved in making working molecules. Much of it is repetitive; that is to say, the same sequences in chemical code occur over and over again in the one cell. In 1969 Roy Britten, then at the Carnegie Institution in Washington, collaborated with Eric Davidson at the Rockefeller University in New York, in working out a theory as to how all this additional DNA might be organised to provide a control mechanism for the genes. Now they have joined forces at the California Institute of Technology and are engaged in a long series of experiments to test the theory.

Britten and Davidson propose that, alongside each 'producer' gene that actually codes for a working molecule, lies another gene that makes nothing, but controls the producer gene. In each chemical process of life a number of genes have to come into play simultaneously. If all of these are accompanied by the same control genes, then a chemical signal which triggers one control gene will trigger all and put the whole set of producer genes in operation at the same time. The triggering material which sets the controls to 'go' would, on this theory, be nucleic acid manufactured by a master gene in multiple copies. The master genes, in their turn, are under the control of chemical signals coming both from within the cell itself and from outside . The whole system of genes is thus subject to the general life of the body and, in a

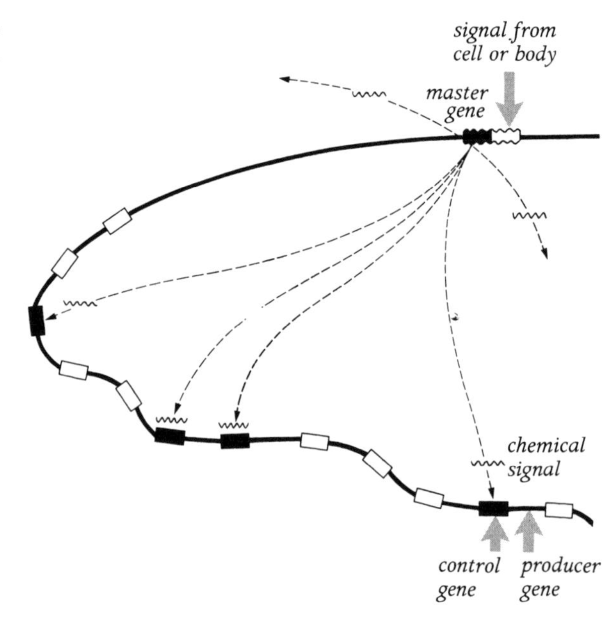

The Britten-Davidson theory of gene regulation. Chemical signals go out from a master gene and match up to corresponding sequences that are repeated many times over in the genetic material. These repetitive sequences act as controls for ordinary producer genes lying alongside them. On receipt of the signal, the control genes activate the producer genes. Thus one master gene can regulate a whole battery of genes.

growing animal, to information about the part each cell is supposed to play.

If the theory is right, then many of the repetitive genes known to exist in cells must be interspersed between the different producer genes. The methods for checking whether they really do occur in that way are difficult and highly technical, but in principle they resemble the children's game called 'farmyard pairs'. In that game each child is told what animal he represents – cow, duck, dog or what you will – and has to rush around mooing or quacking or barking until he finds in the uproar another representative of his species.

In experiments with toads and sea urchins to test the Britten-Davidson theory, the 'pairs' consist of the DNA of the genes on the one hand and their messages or control signals on the other hand – RNA molecules made by the genes. The latter will recombine with their appropriate genes if they encounter them. The repetitive genes are then like children given an unfair advantage because an organiser has told many of them to be dogs; they will pair up more quickly than will the others for which partners are scarcer. The molecular pairs can be picked up because they stick to a mineral called hydroxyapatite, while unpaired nucleic acids are not adsorbed in this way. By seeing how quickly different sections of the genetic material find partners and settle on the hydroxyapatite, the Caltech experimenters are obtaining promising evidence for the presence of repetitive genes among the producer genes, as predicted. But much more work will be needed to clinch the theory.

Britten and Davidson liken the control system of genes to a computer. A computer can be set a completely new task by writing a new programme; it is not necessary to rebuild the machine. In a similar way, the evolution of higher animals may have been a matter of writing new genetic control programmes rather than of evolving new producer genes. Most of the producer genes used by human beings are present in an amoeba; what distinguishes us is the genetic computer that can take an amoeba's genes and print out all the poetry of growth and form and organisation in a complex animal.

After this brief detour to current experiments in developmental biology, it is time to return to those little organisms that were becoming ever more sophisticated towards the end of the Age of Microbes. The reason for the digression was to see what further skills were needed before complex animals could emerge. The answers at present remain imprecise but, as Eric Davidson puts it:

> Consider the appearance of novel structure and function in the course of evolution. Our conception is that a small change in the DNA sequences responsible for the control of the activity of many genes could result in what appear to be huge changes in the characteristics of the organism.

By 600 million years ago the animals possessed:

(1) the basic chemical powers accumulated by bacteria;
(2) the cellular machinery and sexual reproduction associated with cells of the modern type;
(3) the means of controlling the genes and cells throughout a complete body; and
(4) a variety of design principles representing basic strategies for life.

For the Age of Visible Animals to commence, all that was required was that complex organisms should by their internal chemistry make hard shells out of the salts of the sea, and so leave more conspicuous traces for the fossil-hunter to find.

A dragonfly immortalised in stone. Protolindenia wittei *lived in Bavaria about 150 million years ago. It represents the spectacularly successful animals with segmented bodies, the chief design that rivals backbones.*

Chapter 5 Visible Players

The transition from the Age of Microbes to the Age of Visible Animals occurred between 800 and 600 million years ago. The first animals composed of many cells rather than one, and which possessed at least primitive systems of genetic control for regulating growth and function, were equipped to evolve in a spectacular fashion. The microbes had accumulated a vast repertoire of chemical tricks; now was the time for organising them in organs. But how, if you were God, would you set about designing an animal out of a mere collection of cells? In practice, the design principles that the meanest of animals evolved in this period of transition, by chance and natural selection, are so cardinal that what comes later seems like a downhill run by comparison.

A many-celled animal justifies its existence, by the test of evolution, only if it secures food more reliably than individual cells could do. As the food consists primarily of living or dead microscopic plants, drifting in the water or sinking to the bottom, any advantage comes from being able to process a large volume of water. One simple method is to fashion a filter, using cells that make suitably robust material (stony or horny compounds) and use the whip-like flagella to drive water through it. Then you have a sponge. Another way to search water is to wave tentacles around in it. Let the body drift and it can be soft: a jellyfish. Let it be stronger and fix itself to the seabed: a sea anemone or a coral. The flatworm, too, was presumably among the oldest designs for many-celled animals: a strip of tissue which propels itself by means of cilia, to look for food.

But other animals built envelopes that could be filled with water – the first hollow bodies. The initial benefit of the 'coelom', as the envelope is called, was to provide rigidity, much as a toy balloon blown up with air retains a certain shape and strength. The water-filled coelom made the animals robust enough for burrowing into the mud of the seabed to quarry for food of past seasons. In human engineering, the potter's wheel showed little promise of the clocks and turbines that were to derive from it; no more do the comparatively simple structures of the early burrowers hint at the complex organs and complex animals, including ourselves, made possible by a hollow body.

Fanfare for the worms

James Valentine, of the University of California at Davis, has for some years been pondering the revolutionary developments in animals just before the Age of Visible Animals. Developing ideas of other palaeontologists, he thinks that, in a rather short period, the marine burrowers evolved almost every variation in basic design used by the most successful kinds of animals – as well as variations of more modest achievement. Occurring in all the seas of the world today are little animals called peanut worms. With a simple coelom and tentacles around their mouths, they represent one of the designs adopted by marine burrowers; it worked well enough but contained no seeds of further progress. Somewhat more ingenious was a division of the coelom into a few compartments with an orderly row of tentacles at the front end. An animal like this could sit securely in the seabed, with its tentacles protruding to snatch food from the water. From this concept are derived the marine animals called bryozoans and lampshells, which have played an important part in the life of the sea during the past 600 million years.

Two other designs were far more notable. The first of these was the annelid worm, similar to the earthworm of today, being composed of a string of similar segments, each with its own coelom and its own set of organs. A nerve ran the length of the body to co-ordin-

Crucial steps in the evolution of animals. The upper diagrams show, schematically, design principles first embodied in various marine burrowing worms, some of which gave rise to the world's most conspicuous animals (after J. Valentine).

ate the activities of the various segments. It then required only that the segments should develop legs for the annelids to give rise to the great group of animals known as arthropods, which evolved into the insects, spiders and lobsters of today.

The other extremely successful design was the one that led to us. The burrowers in question had only a limited number of compartments in the coelom but they also used the principle of the coelom to make a superior sort of tentacle. Two great groups of animals derived from ancestors like these: the echinoderms, which include starfish and sea urchins; and the vertebrates, from fishes to man. Unlike all the other groups, the vertebrates appeared after the Age of Visible Animals had begun; it is not surprising that the most elaborate of animal designs should have taken longer to emerge.

Only one of the great groups of higher animals does not fit particularly well into Valentine's scheme of design principles from early burrowers. These are the molluscs, such as whelks, snails and octopuses. Like the others, they have hollow bodies, but they are not related to any known form of burrowing worm. In basic design molluscs are a compromise between the peanut worm and the annelids: that is to say, their coelom is undivided but they have repeated systems of organs along their length.

Valentine and his geological colleague at Davis, Eldridge Moores, offer a reason for the rapid evolutionary advances in burrowing worms just before the Age of Visible Animals. It was that, not for the last time, the land of the world was massed into a supercontinent. For the early many-celled animals the most important consequence was an inconstant climate in the coastal waters where they lived. Just as Asia today creates the monsoons, and with them a strong seasonal variation in the growth of marine plants in the surrounding waters, so would the former supercontinent, to an

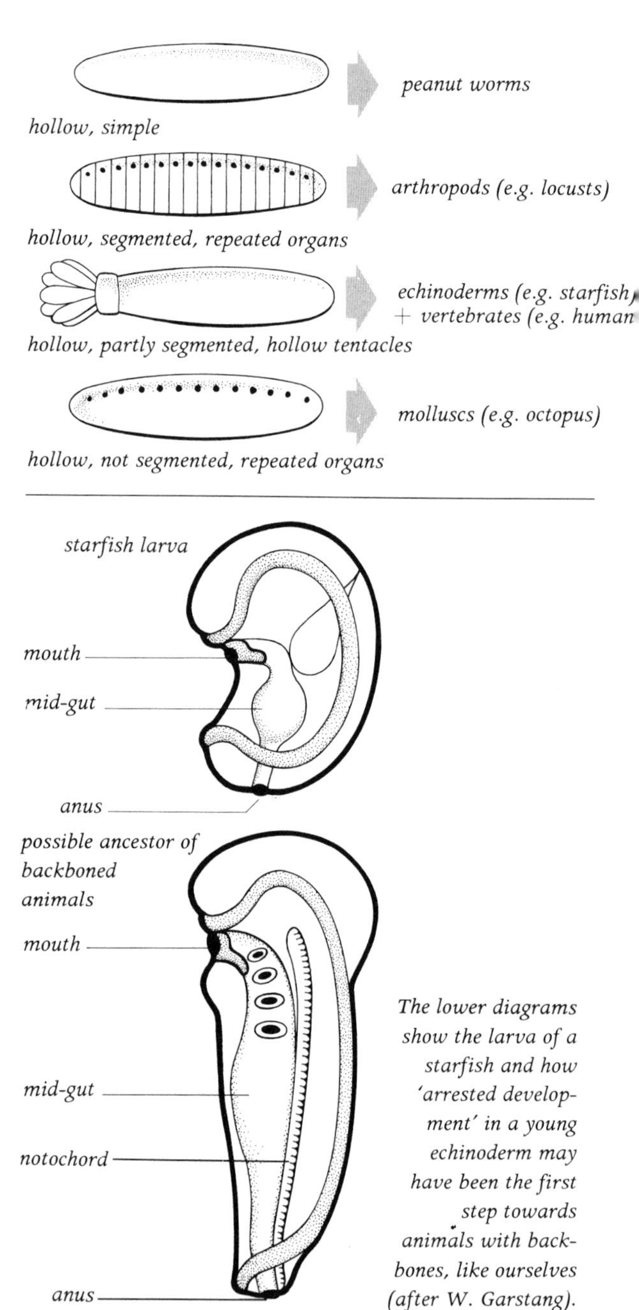

The lower diagrams show the larva of a starfish and how 'arrested development' in a young echinoderm may have been the first step towards animals with backbones, like ourselves (after W. Garstang).

even greater degree, bring long months of diminished food supply for the marine animals. But the remains of plants that were overlooked when they died accumulated in the mud of the seabed. That alternative food supply was not subject to seasonal changes, so to be a burrower brought great advantages.

Coalescence and break-up

Part of the evidence for the existence of a supercontinent about 600 million years ago is in the mountains of Asia, Africa and South America, built around that time. This brings us back to the story that we left aside when, in Chapter 3, the molecular biologist and the radiochemist were finding historical parallels between mountains and the molecules of life. In our imaginary games of 'Doing without fossils' the scientists were able to work out the evolutionary relationships of a variety of present-day species, from pigs to yeast, by matching differences in a particular molecule to the ages of the Earth's mountains. One implication, that molecules evolve at a roughly constant rate, then absorbed our attention. We can now return to the second important aspect. Mountains affect the evolution of species that live near them, of course, but they are not themselves the reason for great evolutionary changes such as the invention of the burrowing worms and, later, of birds. Branchings in the evolutionary tree, which let the molecules diverge and evolve differently, and the principal mountain chains are both effects of a single cause – the movements of continents.

Repeatedly, during the history of the Earth, the land has massed itself into great supercontinents which have then broken up again. This history and the reason for it have become clear only in the past few years, with the discovery that the outer shell of the Earth is broken up into great 'plates'. These plates slide slowly about the face of the globe, growing from the mid-ocean ridges and being destroyed at deep ocean trenches where they dive back into the main body of the Earth. The continents are mere passengers on the plates but are made of light rock which will not sink at the ocean trenches; accordingly their peregrinations about the globe sooner or later bring them into collision with other continents. Many of the major mountain chains of the Earth are simply wreckage thrown into the air by these slow but forceful collisions.

From the evolutionary point of view, there are then two kinds of eras. When supercontinents are breaking up, life is relatively easy on and around the different continental rafts. All sorts of new species and families of species develop. But they are simply variations on the main kinds of plants and animals that existed before the break-up. There is no compelling reason why complete novelties should have any great advantage. It is a time for niche-filling and for trying out new applications of old design principles.

Circumstances are very different when the next supercontinent begins to amass. The climate becomes more seasonal, as mentioned just now in connection with the burrowing worms. Moreover, species of the various continents are brought into direct competition with one another; many are extinguished. A dramatic example was the extermination of many species of rather primitive South American mammals only about two million years ago, when their continent became united to North America and more advanced mammals marched southwards. Coalescence of continents is bad for the diversity of species but good for evolutionary novelty, as a way out of the death trap. Eras of coalescence, marked by mountain-building, are more likely to produce radical changes, like the invention of flight, than the eras of continental break-up.

The biological events signalled by the mountain-building of 460 to 225 million years ago correspond to the construction of the most recent supercontinent, Pangaea. During this time, amphibians and reptiles appeared; mammals and birds came slightly later, but while the supercontinent was still in being. The mountains of the earlier supercontinent (700 to 550 million years ago) match the divergence of the ancestors of insects and vertebrates. In 'Doing without fossils' the diagrams indicated a date of about 1200 million years ago for the branching of plants, animals and fungi. Without fossil evidence, who can tell whether the modernisation of cells and the establishment of the great kingdoms of living species were related to geological events? Even so, our simple molecular game took us into a period of evolution that may be for ever unrewarding to the fossil-hunter.

The connection between mountains and evolution is also less clear-cut in the period 100 million years ago to the present. The mammals, of which pigs and rabbits were present-day examples for 'Doing without fossils', originated during the period of the formation of Pangaea, but they remained inconspicuous until the demise of the dinosaurs. That did not occur until about seventy million years ago, when the break-up of Pangaea was already far advanced. Already geographical dispersal had encouraged different lines of mammals to develop independently. The biological difference between pig and rabbit is really rather slight compared, say, with the difference between pig and frog, and their divergence does not represent a major branch of evolution.

We live in a period of continental scattering, with a greater variety of plants and animals than has ever previously existed on Earth. Of the young mountains, the Andes or the Rockies do not represent major continental collisions. An embryonic supercontinent is beginning to form, with Africa and India pressing into Eurasia (making the Alps and the Himalayas) but so far dispersal exceeds coalescence. Fossil-hunters of the far future, for whom our blue whale will be the prize trophy, will find nothing radically new beginning in our time, until man starts inventing completely novel kinds of organisms – but that has not yet happened.

Planetary scene-shifting

Descriptions of the course of play at various stages of the life game can now blend the classical conclusions of fossil-hunters with deductions about the shifting stage on which the organisms were living and evolving, as continents marched about the globe. The idea of relating fossil evidence to continental movements certainly is not new. Until just a few years ago, when the theory of continental drift was still widely regarded as a crazy idea, some of the most powerful arguments in its favour came from the kinship between fossil plants and animals in continents now widely separated by oceans.

Now that continental drift is accepted, in its present guise of plate tectonics, fossil-hunters and earth scientists can join forces and provide mutual checks on their theories. The tracks of continents have to be inferred by judicious pooling of evidence, including: the direction of magnetisation of rocks; volcanic activity associated with the formation or decay of ocean basins; sediments formed or trapped alongside the shifting land masses; climatic changes marked by both the physical condition of the rocks and the character of the fossils they contain; similarities and differences between organisms which tell of connection and separation of land masses.

The biologist now obtains from the earth scientist a wonderful illumination of the history of groups of

Evolutionary relationships among the main groups of organisms. Biologists have long been familiar with the main features but the dates and some of the groupings reflect modern knowledge.

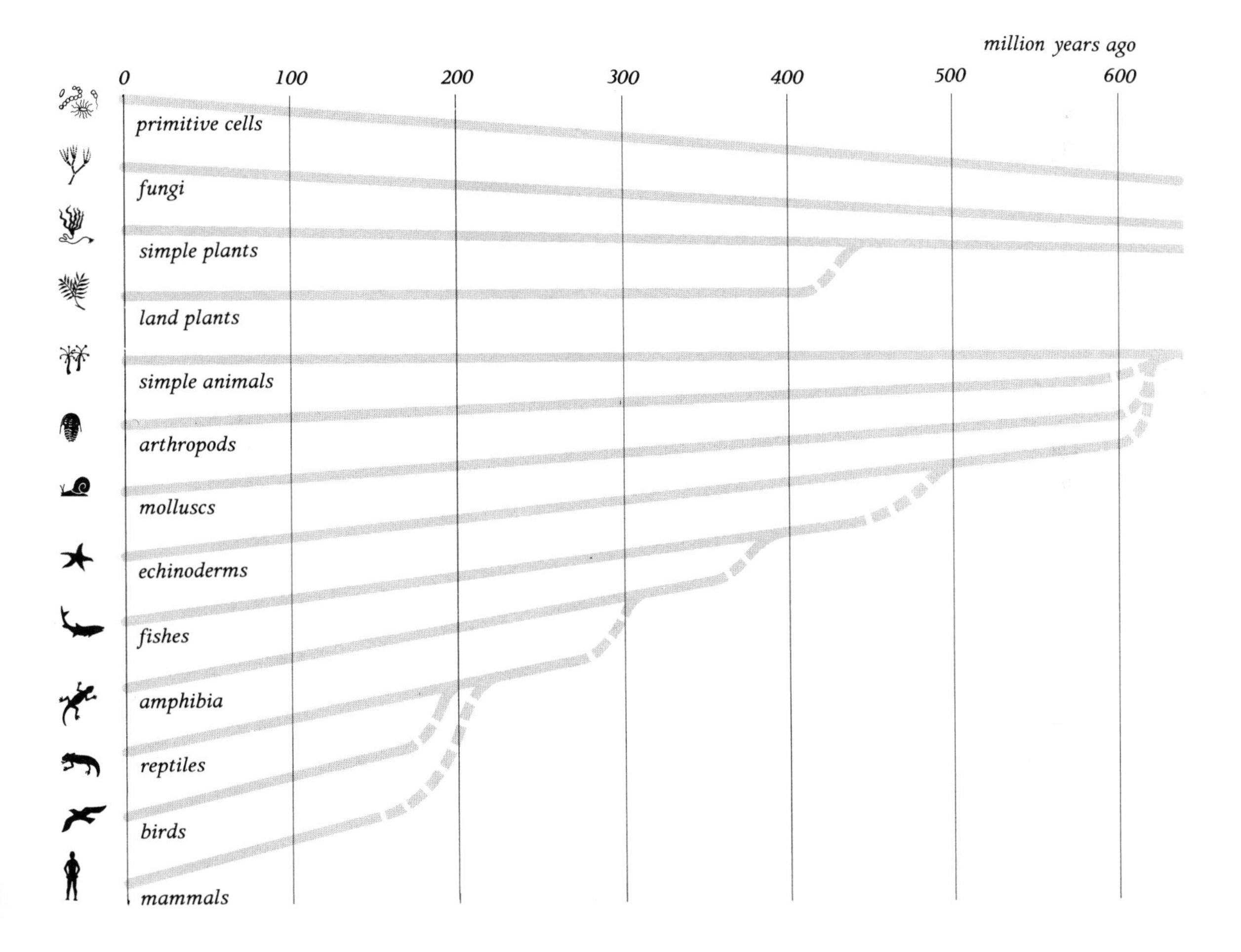

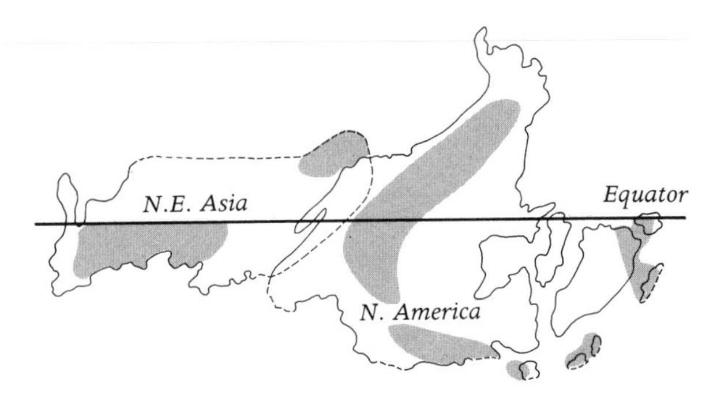

A young kangaroo (right) attached to the nipple in its mother's pouch, at an early stage of development when, in more 'modern' mammals, the mother can still feed the baby inside her.

The diagram (left) shows how small animals, trilobites of a distinctive kind, tie eastern Siberia to North America 500 million years ago. (After H. B. Whittington and C. P. Hughes.)

Occurrences of bathyurid trilobites.

animals. The strange mammals of Australia are a case in point. Australia used to be a part of the supercontinent of Pangaea and shared its common pool of plants and animals. More than 100 million years ago, at an awkward stage in the evolution of the mammals, Antarctica and Australia broke from Africa. Mammals already existed but the placenta had not yet been invented, as the link by which most mammalian mothers feed the baby inside them until it is quite large. By separating when it did, Australia missed this invention; it had marsupials instead. Marsupials give birth when the young are still embryonic. The youngster has to find his way to the nipple and remain constantly attached to it until he is very much bigger. The chief function of the kangaroo's pouch is to protect the baby during this period. The system worked well and a rich variety of mammals with pouches evolved in Australia – until man introduced mammals of more advanced design, which even today threaten the extinction of the marsupials.

The biologist can prevent errors on the part of the earth scientist. Anyone wanting (as some have done) to detach India from Africa nearly 200 million years ago is corrected by the biologist who can point to dinosaurs that India and Africa had in common about 80 million years ago. Again, the earth scientist for convenience treats continental masses in the past as if they corresponded at least roughly to the present day outlines of continents. But when the evidence of extinct groups of animals is mapped on the distribution of continents deduced for 500 million years ago, some revealing discrepancies show up.

In different parts of the world a celebrated group of early animals, the trilobites, evolved in somewhat different ways, creating what are known as faunal provinces. In the simple maps of the earth scientists, Asia and North America appear as separate continents of more or less their present configurations. But in fact there is close affinity between the trilobites of eastern Siberia and those of North America. This encourages the idea, for which there is other evidence, that 500 million years ago eastern Siberia was not a part of Asia but was joined to North America.

The confirmation of continental drift, and especially the exploration of continental movements of more than 200 million years ago, constitute so recent a revolution in the study of the Earth's history that examples like this re-examination of the trilobites are rather scarce so far. Great advances in understanding the events of evolution can be expected as earth scientists and biologists proceed in this exciting review of facts, known for many years, but only now beginning to make sense.

A partial record

Since 1796, when the French naturalist Georges Cuvier astounded his compatriots by unearthing elephants in Paris, fossil-hunters have amassed and described the remains of more than 100,000 different animals and plants that formerly populated the Earth. New forms are still turning up at a rate of many hundreds each year – an indication of how much remains to be discovered. One estimate puts the number of different forms still lurking in the rocks at ten million; that is after allowing for the fact that most living things were unlikely to leave any fossil record.

An animal seeking eventual enshrinement in a museum would be well advised to live in or near the water. To leave an impression in newly forming rock, or to turn into a piece of stone before it rots or breaks apart, it has to perish in peculiar conditions. Soft-bodied plants and animals are much less likely to leave visible remains than those with robust shells or skeletons, although X-rays may in future help to reveal

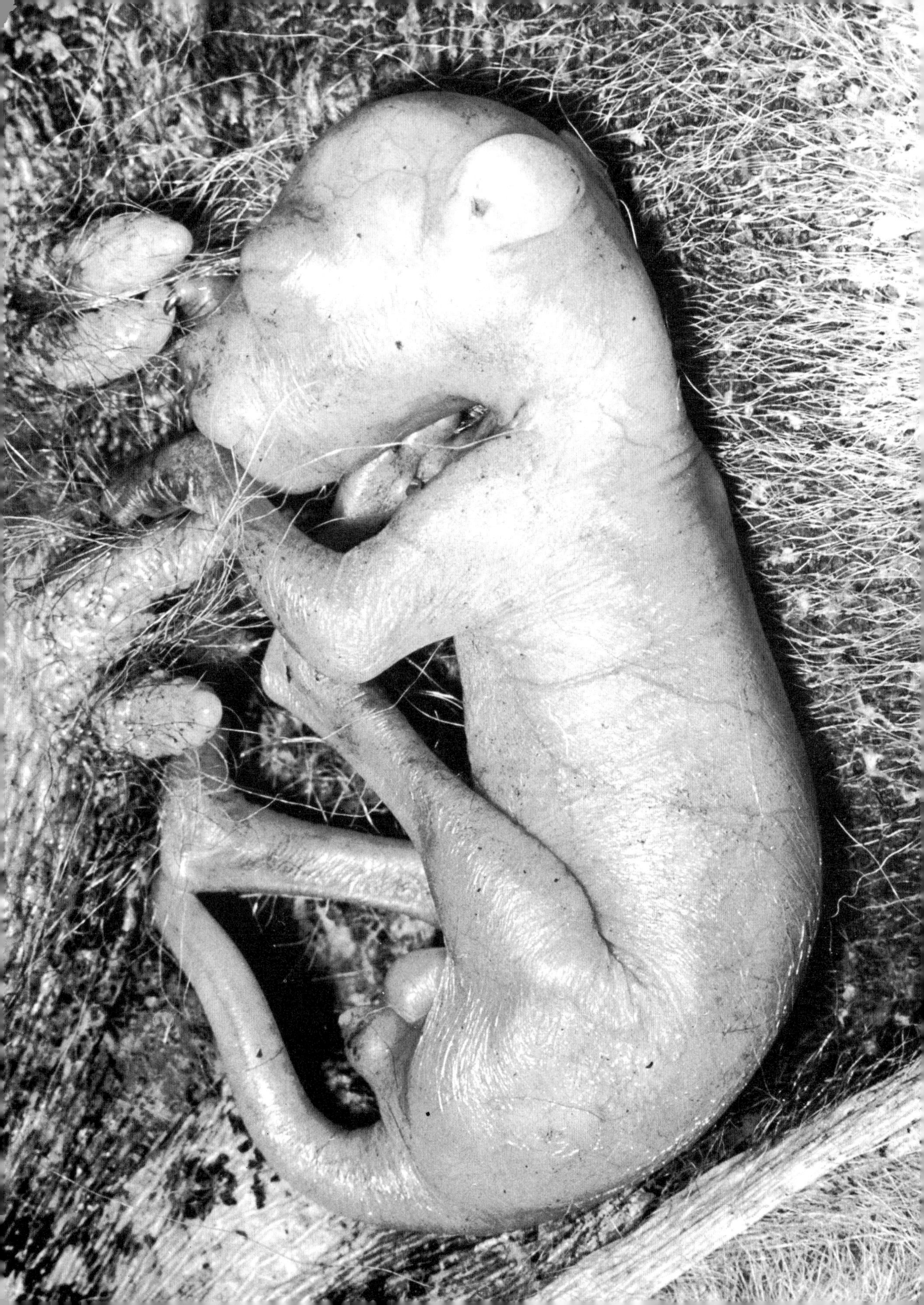

Footprints left in Texas 120 million years ago by a big dinosaur that was being pursued by a three-toed carnivorous dinosaur.

traces of soft tissues in rocks that, to the naked eye, are barren. Insects must have abounded for the past 400 million years but, unless a fly was trapped in amber, its chances of fossilisation were slight indeed. Animals and plants of the uplands, woodlands and deserts are poorly represented among the fossils. The best candidates for preservation are those trapped and killed in floods, quicksands or swamps, but subsequent erosion, or burial at inaccessible depths in the Earth's crust, has eliminated many of them.

Footprints of dinosaurs, always a fascinating sight, are one way in which the fossil record makes up for failures to preserve the animals themselves. Tracks and burrows, even of small animals, preserved in the rocks give insight into the behaviour of animals that lived long ago. The epoch-making arrival of marine burrowers on the scene is recorded more plainly by the animals' activity than by their remains. Very old sediments that have been preserved relatively undisturbed by later upheavals form very neat horizontal layers until, shortly before the Age of Visible Animals, the layers are churned up, indicating the evolution of the burrowing worms.

Charles Darwin wrote lovingly of the fossil bones which 'tell their story of former times with an almost living tongue', but he devoted a chapter of *The Origin of Species* to the imperfection of the geological record. What is remarkable is that, although fossil-hunters have probably sampled less than one per cent of the beings that existed even during the past 600 million years, they can nevertheless give a consistent and plausible account of the main lines of evolution leading to present-day species, and of the principal forms that arose only to be extinguished again. In fact the failure to unearth many prehistoric forms is probably less misleading than the habits of thought about evolution that the available fossils engender.

Our imaginations cannot cope with the great periods of time; nor are we able to visualise the vast numbers of individual organisms, as opposed to species or families of species, that have lived on the planet. Yet the fate of individuals determined the course of evolution. Everyone uses diagrams like branching trees or railway marshalling yards to show how new kinds of plants or animals budded from ancestral lines, creating the present diversity. Such diagrams fit poorly with modern evolutionary theory based on studies of living species. The steady progress they imply is the chief source of false notions of purpose or direction in evolution. The descent of man, for instance, is all too easily seen as advancement through early fishes, to amphibians, to reptiles, to early mammals, to primates and thence to man, down the railway track.

Each 'track' or 'branch' in evolutionary diagrams is really composed of thousands of minor branches, most of which are quickly terminated – a strand of wool would be a better analogy. Each fibre consists of many individuals leading practical lives, in which the course of events determines whether and with whom they will mate and whether the offspring carrying their genes will survive. If we could see a family album of our animal predecessors, going all the way back to the fishes and worms, they would be neither a straight line of single specimens nor the complete panoply of the animal world. Instead they would consist of substantial populations of animals which were nevertheless a very small minority of the whole.

The following pages show a series of snapshots of the Earth and some of its inhabitants at various stages of the Age of Visible Animals. The positions of continents are those worked out jointly by Alan Gilbert Smith and Gill Drewry of Cambridge University and James Briden of Leeds University.

A sea scorpion, or eurypterid, of more than 400 million years ago.

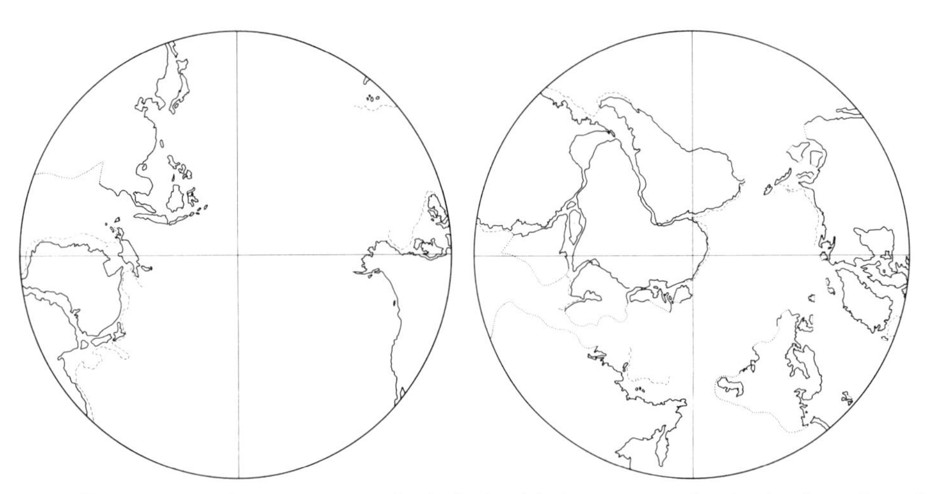

500 million years ago *the supercontinent that had existed during the transition from the Age of Microbes to the Age of Visible Animals was breaking up. Life was confined to the sea and to fresh water; the land was drenched in deadly ultra-violet rays from the Sun. But plants were steadily increasing the amount of oxygen in the air, which in turn was forming ozone high in the atmosphere and beginning to filter out the ultra-violet.*

The most abundant animals of note were the trilobites, so called because of their three-lobed bodies (see photograph facing page 112). They were early members of that exceptionally important group of animals, the arthropods, with jointed legs and segmented bodies. Growing to prominence, too, were the echinoderms, with no plain heads and with the five-sided symmetry still obvious in starfish.

The tadpoles of some echinoderms, or near relatives, 'froze' in their development, producing a sort of boneless fish (see diagram on page 102). But at that time, too, the arthropods included dangerous sea scorpions (see opposite). They made many a good meal of these newcomers and proto-fishes appeared with an armoured back of bone – the origin of the backbone.

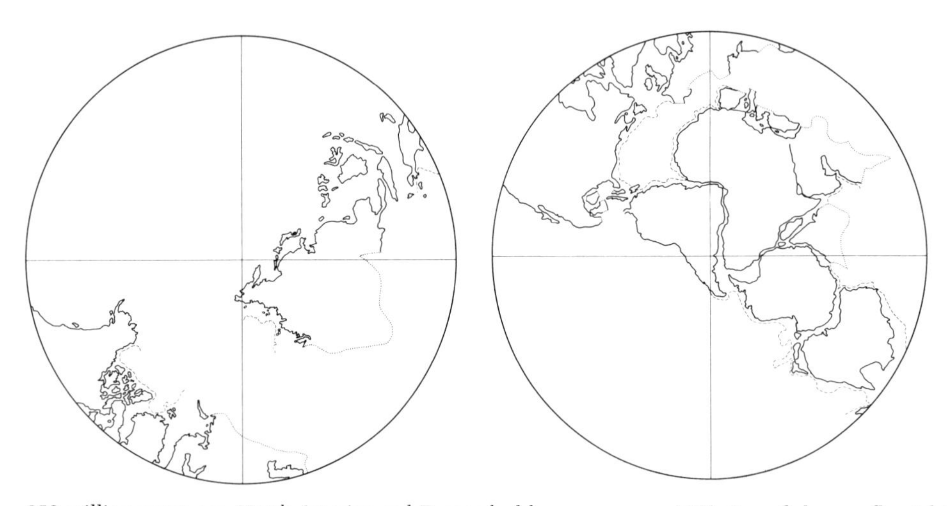

350 million years ago *North America and Europe had become welded together and they were beginning a slow collision with Africa. Plants had invaded the land almost 100 million years before. Their evolution was now in full swing and plants related to the modest club mosses, horsetails and ferns of our own time founded forests of tall trees. Following the plants ashore were the arthropods: scorpions, spiders and the first winged insects.*

In the water, the ammonites had made their debut – those jet-propelled molluscs whose coiled shells are almost as familiar to amateur fossil-hunters as to professionals (see photograph on page 113). Bony fish were flourishing. Some of them developed lungs which could save a stranded fish's life; an ability to slither or crawl in search of another pool was a further aid to survival. The conversions thus encouraged, of gills into lungs and fins into feet, made the first amphibians.

In the words of a leading American analyst of fossils, Alfred Romer: 'To a variety of forms, ranging from insects to other vertebrates, eggs in a pond are desirable amphibian caviar.' Some amphibians, while still finding their food in the water, invented the land egg and became the first reptiles.

The trilobite (right) of nearly 600 million years ago represents one of the great groups of animals formerly inhabiting the Earth. The jawless armoured fish (below) is about 400 million years old. It probably filtered its food from the mud.

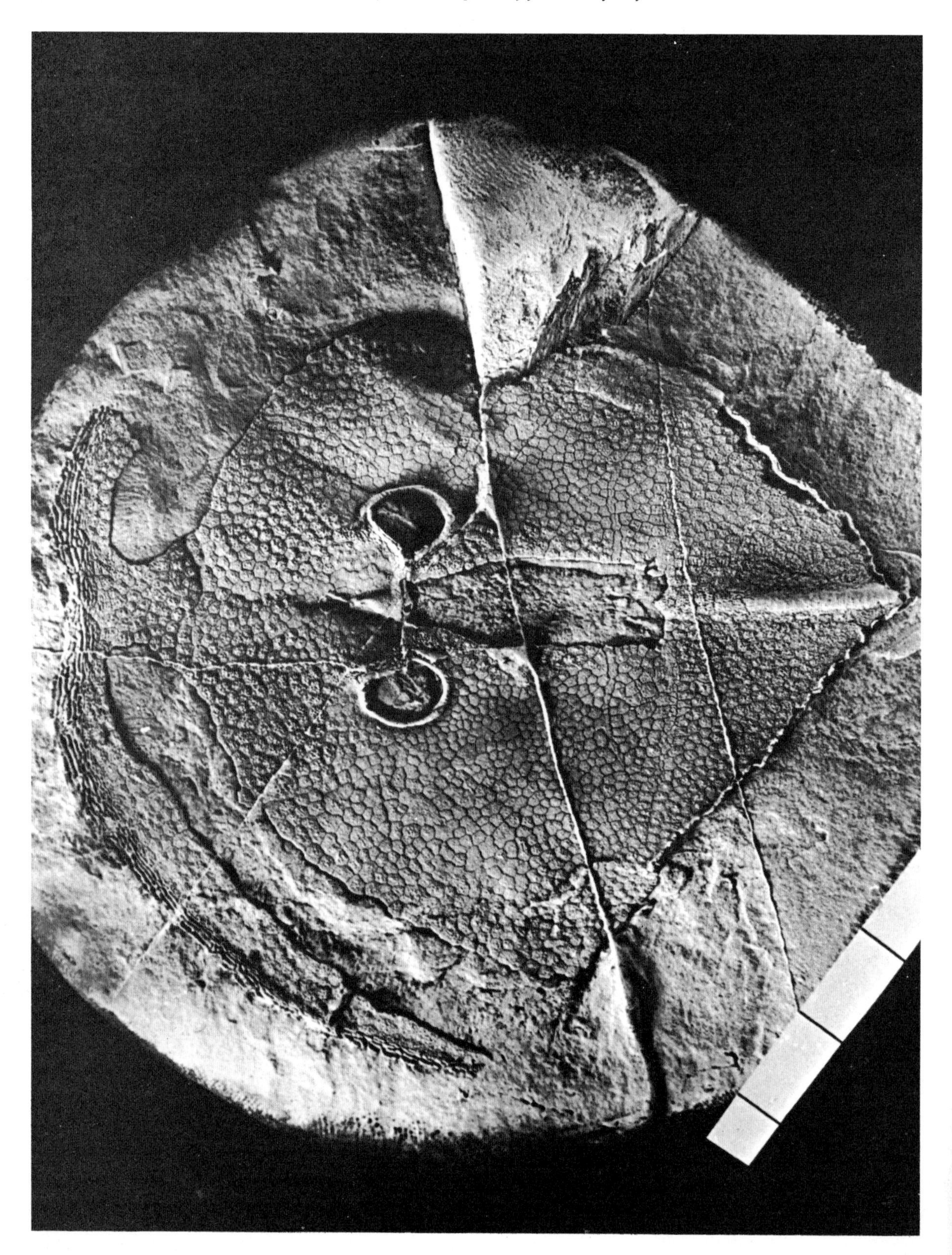

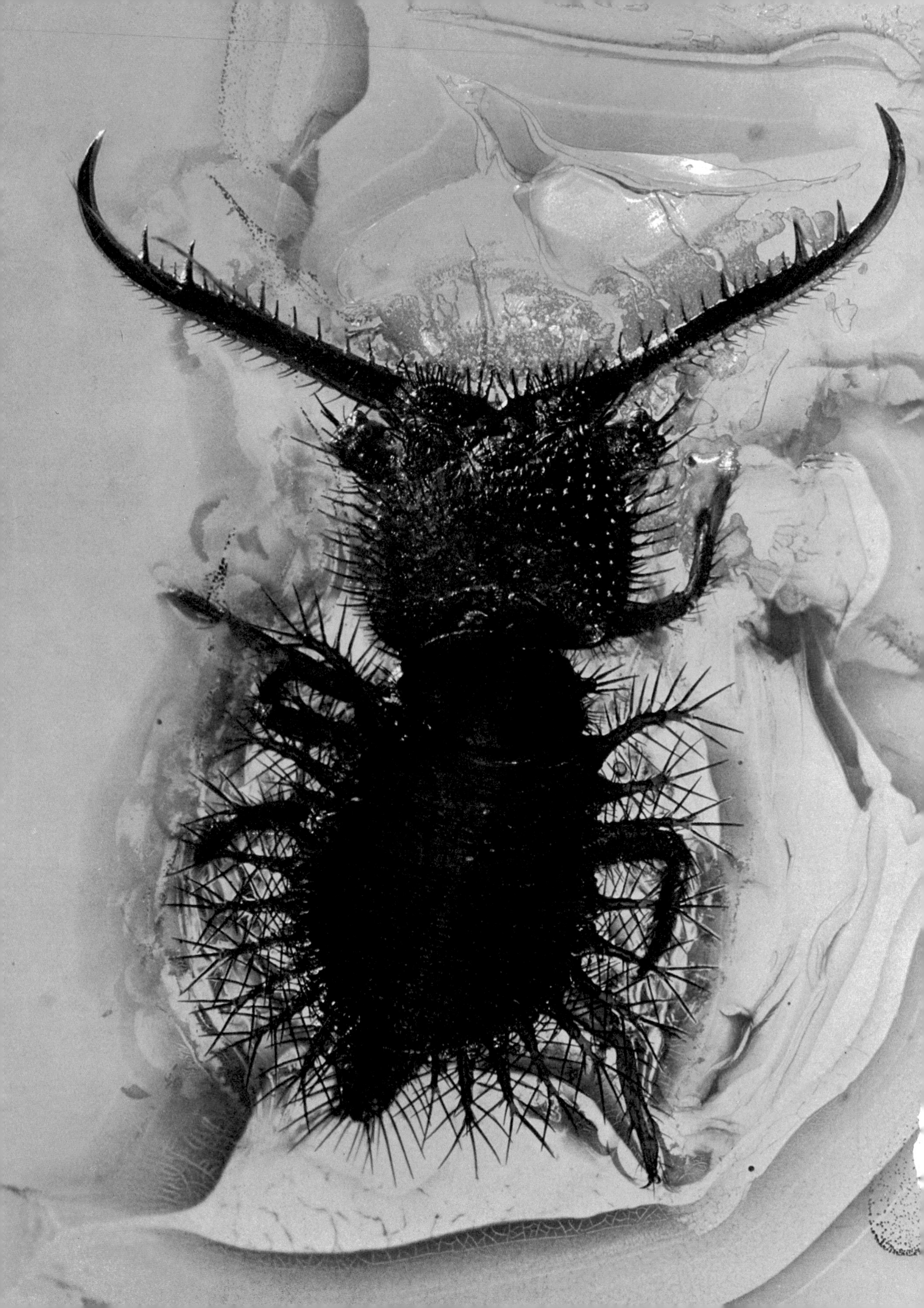

The ammonite (above) was bitten by a big python-like reptile. Both the ammonite and the chief reptiles that assailed them became extinct 70 million years ago in the revolutionary upheaval in which the dinosaurs died out. The insect preserved in amber (left) lived about 30 million years ago. It is the larva of a flying insect related to the alder-flies of today.

The primates were recognisable 50 million years ago – among them this lemur-like fossil skull found in Europe.

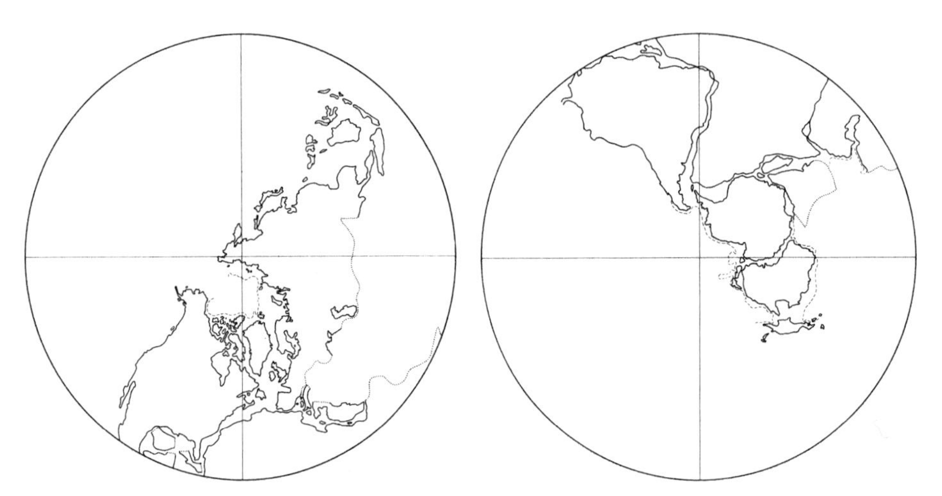

By 200 million years ago, *the continents had merged into a huge land mass – Pangaea. The drastic alteration in the global environment caused a massacre, known to fossil-hunters as the 'Permo-Triassic crisis', in which great armies of animals perished. Trilobites, sea scorpions and the old-fashioned corals were exterminated and almost every group of animals suffered heavy losses. Three-quarters of all families of amphibians disappeared and even more of the newly evolved reptile lines were wiped out. The plants, as far as anyone can tell, continued undisturbed by the great changes.*

In the aftermath, though, the most spectacular developments were among the reptiles. The dinosaurs were the undisputed masters of Pangaea and reptilian sea monsters waged war on fishes. But the first mammals appeared at this time, resembling present-day shrews, both in size and in their diet of insects. Both mammals and birds evolved from reptiles that had a hard time competing with dinosaurs. Little shrews that Tyrannosaurus *would have disdained to notice embodied the inventions like nipples, fur and big brains that were to supersede the great reptilian experiment.*

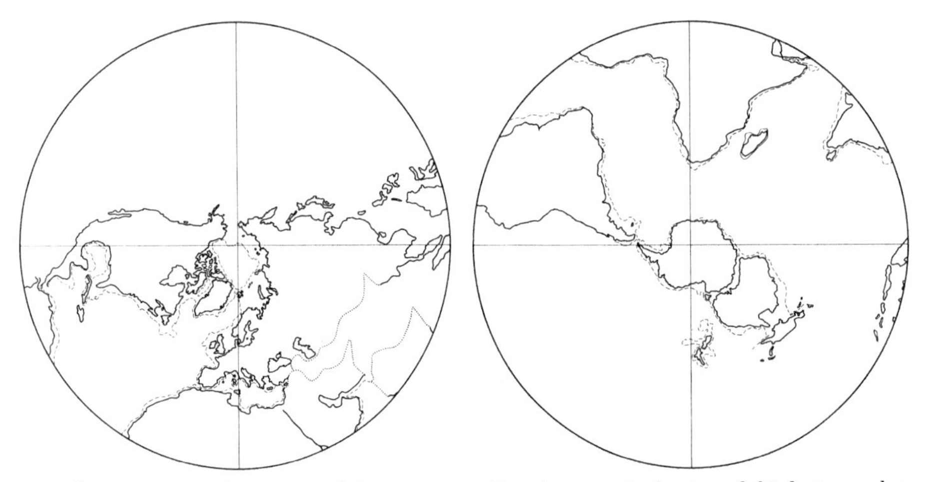

50 million years ago, *the continental theme was one of break-up. The dinosaurs were all dead; so were the ammonites and other marine animals characteristic of the era that ended abruptly about 70 million years ago. One theory is that so much chalk was laid down in the latter stages of the dinosaurs' reign that enough carbon dioxide was removed from the air to cool the Earth; the naked dinosaurs died of cold but furry mammals survived. The many ways of life vacated by the dinosaurs were taken up by mammals. But the mammals were not mere understudies; they brought to the scene agility and brain-power never seen before.*

At the time of this last snapshot a group of animals was emerging with clever hands and feet and special investments in eyes and brains – the primates (see photograph opposite).
Yet no evolutionary leap has occurred in the last 50 million years of the Earth's history to compare with the invention of land eggs, lungs, warm blood, nipples or wings – unless it be the invention of speech in humans. Even among the mammals, the fastest evolving group – even for man – subsequent evolution has been mainly a matter of adapting and refining equipment already available.

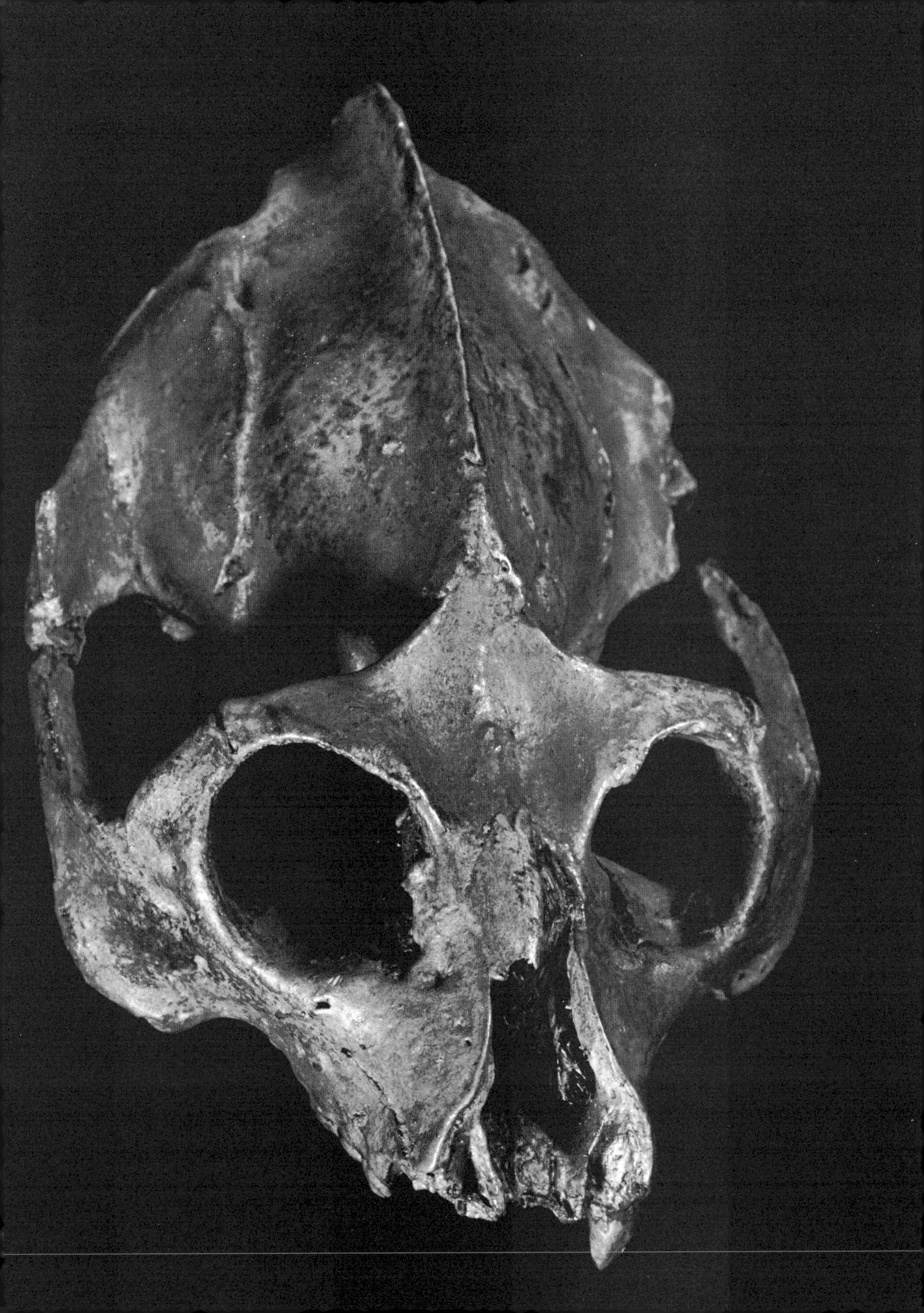

A lemur, Indri indri, *in the forest of Madagascar. It represents the primates that evolved separately in that island.*

Country cousins

Looking for lemurs is easier at night. What you do, in the forests of Madagascar, is to hold a powerful torch alongside your own eyes and scan the trees with it; before long you will see a pair of orange, strongly reflecting orbs staring widely at you from among the leaves. Animals that would be hard to spot in daylight give themselves away by their reflecting eyes. By day, you scramble and sweat through the tropical forest, following the yelps. With an expert guide, and if you are sufficiently quick, silent and lucky, you may at last be rewarded with the sight of *Indri indri,* the mother minding the baby while the father sits farther along the branch watching anxiously for intruders like yourself. Lemurs come in all sizes, from species as small as mice to big ones like *Indri indri,* and they occur only in Madagascar. Continental drift removed that island from Africa, taking with it some of the early primates, which have since pursued a course of evolution quite independently of monkeys and apes in other parts of the world.

Our instructor in the Malagasy forests was a young research student from London, leading a hermit-like existence recording the activities of various families and groups. Lemurs interest the investigators of animal behaviour because of the wide range of social organisation they show, from solitary animals to large troops. For primates, including ourselves, more than for any other animals, evolution has been a subtle interplay of the talents made possible by an enlarging brain, whether for well-controlled agility or durable social organisation. But the accidents of changeable geography have also played a decisive part. Alterations to the physical and climatic maps of the world have separated the primates into different areas and dislodged some of them from the trees.

Most species of primates still live in trees as did the ancestors of all present ground-dwelling monkeys, apes and men. The various designs of primate bodies correspond to the different ways of moving around that have evolved among them: clinging and leaping, clinging and swinging, four-footed walking, or two-footed walking. The ability to stand on two feet leaves the hands free for other purposes – throwing, for instance. Man's nearest relatives, the chimpanzees, can throw after a fashion.

Leave a stuffed leopard sitting where free-living chimpanzees will find it and they will, after an initial retreat, join forces to harass it. They charge the leopard and throw branches at it. Even when they tumble to the fact that it is not alive, they may continue with highly emotional displays around the dummy. This is the favourite experiment of Adriaan Kortlandt of the University of Amsterdam. In several expeditions to West Africa, he has used the stuffed leopard to test, among other things, the way chimpanzees throw.

All chimpanzees are poor throwers, but Kortlandt finds striking variations in the action. Forest-dwelling chimpanzees use only a high overarm throw, like that of a human two-year-old. They fling branches downwards – appropriate enough in an animal accustomed to retreating up a tree when danger lurks, but they do the same on the ground. Chimpanzees living in more open country use an underarm throw, which a three-year-old human child can accomplish and which we often describe as 'girlish'. Thus the animals achieve a more horizontal trajectory. But the forest-dwellers, having little scope for it in the thick foliage, have evidently lost the capacity for underarm throwing.

The shoulders and arms of chimpanzee and man are very alike, yet the human capacity for throwing hard and straight is greatly superior. In particular, older humans use a low overarm throw, with the hand close to the shoulder. This action, suitable for stabbing

Adriaan Kortlandt demonstrates the low, over-the-shoulder arm movement, as in stabbing or spear-throwing, that is characteristic of man. In one of his research films (right) a chimpanzee is wielding a stick overarm as it approaches a stuffed leopard.

and spear-throwing, was a crucial innovation according to Kortlandt; it is a relic of the time when early man 'took his canines in his hand'. Once armed with a stabbing stone or spear, man could afford to evolve the more even teeth that are one of the hallmarks distinguishing us from the apes. Other changes were concomitant. You cannot be much good with a spear if you walk around on your knuckles as the apes do. Apes can stand up when occasion demands, but lacking the long, well-hinged legs of man they walk awkwardly. An upright posture, besides helping early man to wield his spear, extended his range because he could more easily carry his water supply in the form of fruit.

Baby apes, unlike their parents, look almost human. This is one of the facts of life which shout most loudly about the evolutionary importance of the systems that regulate growth and development. Evolving from ape-like ancestors, we have preserved into adulthood characteristics of the infant ape, in the lack of hair, in the proportions of legs and trunk, and in the shape of the skull. But why should man have thus diverged from the apes? And when did the various partings of the ways occur, in the evolution from early primates to man?

The search for relationships among the fossil ancestors of apes and man has vexed students of human origins for a hundred years. Each new find has seemed as much a confusion as a clarification, as each researcher draws up his own scheme for primate history. Into this professional confusion biochemists recently rushed, offering differences between molecules as a way of dating the divergences of one group of primates from another. Uncertainties about how constant the rate of molecular evolution has been in the primate era have led, though, only to an intensification of the controversies.

Rifts among the apes

Early in 1973, Adriaan Kortlandt offered a new clarification of the evolution of apes and man. It was based neither on molecules nor on the traditional haggling over teeth and pieces of skulls, but on the new geology of drifting continents. The cardinal lesson from studies of present-day evolution is that, among higher animals, new species occur when groups of related animals are physically separated so that sexual relations between them cease. In and around Africa, the evolutionary heartland of the primates, comparatively small geographical changes have had far-reaching consequences of this kind. Arabia, appended to Africa, united with Asia when a former ocean closed; then the Red Sea opened to divide Arabia from Africa. Associated deformation of Africa created the two great rift valleys that lie towards the eastern edge of the continent.

This was the shifting ground upon which the drama of the evolution of apes and man was enacted. And for Kortlandt the dates of these events, which divided the early apes and let them evolve in their different ways, provide the time-markers for the events. The creation of new lakes and rivers, and climatic changes which made deserts, are important geological details to be added. So is the salient fact that apes cannot swim; even if their ancestors had tried, the crocodiles would have had them. Perfect geographical barriers are not required for this formation of new species – just very strong discouragements to travellers, and hence to the flow of genes between different populations. Here, in outline, is the story that Kortlandt now tells.

About twenty-five million years ago ancestral apes were living in various parts of Africa and were spreading into Asia and Europe. They were tree-dwelling; they had ape-like teeth but otherwise were more like monkeys. With the opening of the Red Sea and associated flooding of Suez, some of the primitive apes were cut off from their relations in Africa and evolved in their own way. In Asia they produced the orang-utan of the present day. The orang-utan differs from other modern apes in being completely adapted to forest life. With no hint of anatomy or behaviour suited to living on the ground, this animal represents the most advanced form of life attained in the trees.

In Africa and Europe the early apes evolved feet and other features suggesting that they spent part of their life at least on the ground. They were very adaptable animals, living in all sorts of environments. In Europe, though, the bears arose and that, according to Kortlandt, was the most likely reason for the eventual extinction of the European apes. In central and southern Africa the apes were protected; bears never managed to cross the barrier created by the Sahara Desert. Even horses took six million years to make their way south of the Sahara.

The next big event for the African apes began twenty million years ago with the appearance of great cracks across the continent, the rift valleys. After a few million years of volcanic activity and irregular geological movements which must have confused the apes but not divided them, the newly sunken floors of the rift valleys became flooded. By fifteen million years ago a continuous chain of lakes and rivers, teeming with crocodiles, reached from Lake Tanganyika to the Nile and, in the south, from Lake Malawi to the Zambesi. The ancestral apes were divided once more.

The apes living to the west of the rifts evolved into gorillas and chimpanzees – the gorillas primarily in the forests and the chimpanzees in somewhat more open country. To this day chimpanzees are confined to a segment of Africa marked out by the Western Rift Valley in the east, the Sahara in the north, and the Congo River in the south – although a separate species of pygmy chimpanzees lives to the south of the Congo.

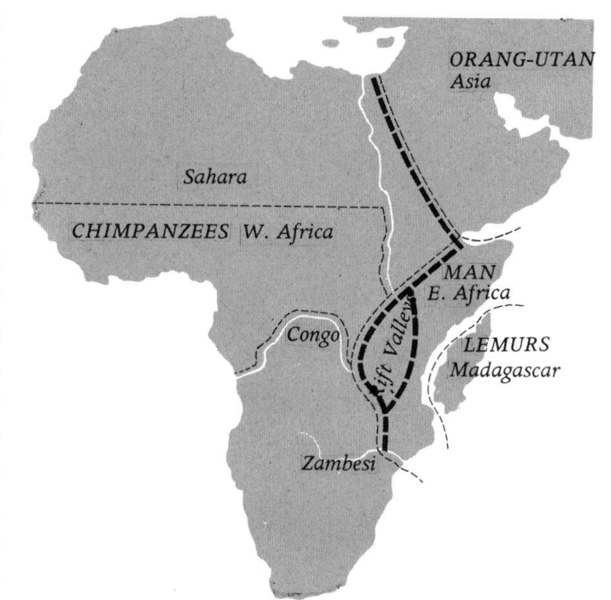

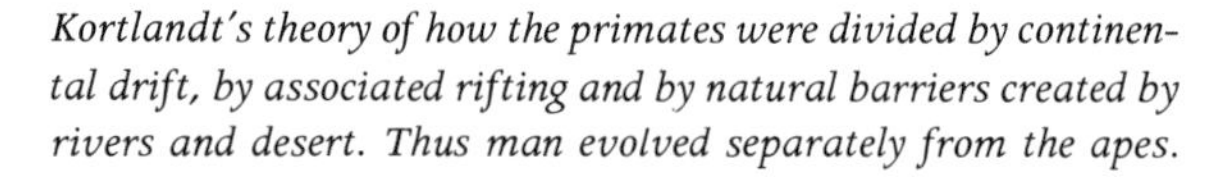
Kortlandt's theory of how the primates were divided by continental drift, by associated rifting and by natural barriers created by rivers and desert. Thus man evolved separately from the apes.

No great apes ventured back into East Africa until a few thousand years ago when new volcanic eruptions created a natural causeway across the Western Rift.

'East Africa was virtually sealed off,' Kortlandt says, and this is his explanation of the separate evolution of some of the ancestral apes that led to man. The region where they lived, Tanzania, Kenya and southern Ethiopia, became drier and less thickly wooded. In this changing environment, over fifteen million years, certain improvements to the legs, refinements of the hands and of their co-ordination with the eyes, and an enlargement of the brain, produced an animal quite distinct from the other primates. He became chemically dependent on meat and administratively dependent on his parents and comrades: a clever, sociable hunter. Unfortunately for present-day investigators, he also had the wit to avoid being drowned in marshes and quicksands, which are the fruitful sources of animal fossils. As a result, tracing the course of his evolution in East Africa has been as much a matter of opinion as of factual evidence.

Leakey's luck

When the young Kenyan hunter of human predecessors, Richard Leakey, came to London in November 1972 to announce an important new discovery, he was excited enough, but rather put out about the nature of his find. It did not suit the style of research he wanted to cultivate. He explained to me that he sought a different kind of reputation from that of his celebrated parents, Louis and Mary Leakey. They had repeatedly astounded their anthropological colleagues with finds, at Olduvai Gorge in Tanzania, of individual fragments of skulls and other bones that compelled fresh thought. 'Leakey's luck' was notorious. But modern evolutionary science demands more than that; evolution is no longer to be seen as one precisely defined type of animal giving way to new types, but as a process involving interbreeding populations in which certain genes gradually become more common. So when Richard Leakey, formerly an organiser of safaris for tourists, was seized with the urge to look for fossils in the semi-desert of Lake Rudolf, he resolved to work differently. He would concern himself not with individual prize exhibits but with collecting many fossils of the same population. Only in that way could reliable, statistical conclusions be drawn from them.

A great virtue of the East Rudolf sites, which Richard Leakey had selected from the air, was that they yielded fossils by the dozen. The area had also been dusted from time to time by volcanic eruptions, leaving rock-layers that could be dated with precision. They helped firmly to establish, for one thing, the co-existence of two man-like animals near Lake Rudolf a couple of million years ago, 'Homo' (though not yet like us) and 'Australopithecus'. Two but not, as some investigators would have said, three man-like animals, for the finds seemed to clear up a problem that had confounded evolutionists for many years, about two sorts of Australopithecus.

Australopithecus was a man-like ape abundant in the south and east of Africa a million or two years ago. Its fossils had been classified into two distinct species, one bigger than the other. A lot of learned discussion ensued about how these two species were related to each other and about how one, or the other, or both, might have been related to man. Richard Leakey and his colleagues collected remains of about fifty Australopithecus individuals from East Rudolf. From careful measurement of individuals known to have been living in the same place, they concluded that, far from being different species, the bigger ones were the males and the smaller ones the females of the same species.

Skull 1470. This broken skull was found by one of Richard Leakey's collectors, Bernard Ngeneo, in rocks of East Rudolf, now said to be almost three million years old. It is remarkable because of its relatively large size.

One of the marks of man is the lack of much difference in build between the sexes. If correct, a contrast between the Australopithecus males and females is a very apish character for allegedly recent ancestors of ours. Associated with the Homo remains in East Rudolf were signs of man-like social behaviour with organised settlements, fairly sophisticated tool-making, and a hunting way of life.

Up to 1972, then, Leakey's statistical approach to the field work, which he conducted as director of the National Museums of Kenya, was paying off well. His results meshed well with those of French and American fossil-hunters working in the Omo region of Ethiopia, just to the north of Lake Rudolf. But then 'Leakey's luck' was visited on the son. A single broken skull, no. 1470, turning up at one of the East Rudolf sites mocked the careful statistics and raised doubts about pre-existing theories of human evolution.

By a typical version of the human lineage, as offered *circa* 1972, the ancestral ape begat early Australopithecus; early Australopithecus begat the most primitive Homo (as found at Olduvai and at East Rudolf, too, before no. 1470); primitive Homo begat *Homo erectus,* a species of man that spread all over the Old World – Pekin man and Java man are famous examples. A quarter of a million years ago men that can just about be classified as *Homo sapiens,* but still very different from ourselves, began to emerge. They evolved first into the Neanderthalers of the receding foreheads and finally, about 50,000 years ago, into modern man, *Homo sapiens sapiens.*

Skull no. 1470 bursts through this scheme, in the vicinity of 'Australopithecus begat Homo begat *Homo erectus*'. The skull is nearly three million years old and the pieces assemble into a rounded cranium about 800 cc (cubic centimetres) in capacity. That is significantly bigger than the skulls of the other Homo, living later by Lake Rudolf, and nearly twice as big as the Australopithecus of only one million years ago. Modern man has a cranial capacity of about 1400 cc. In its size, roundness and other characters, skull 1470 seems more like us than do more recent fossils held to represent our ancestors.

Anthropologists are wondering what to make of it. At the time of writing, there seems to be little reason for doubting either the age or the size of the reassembled skull. What is in question is its meaning. The conservatives preserve the old scheme with the least possible modification by declaring that 1470 is just a big Australopithecus. More radical interpretations expel the Olduvai Homo from among our forefathers. In that case 1470 is either a direct forerunner of *Homo erectus* or perhaps – in view of its remarkably modern appearance – the start of a lineage that bypassed even *Homo erectus* on the long road to *Homo sapiens.*

Yet Leakey hopes that a lesson of 1470 will be the doubts it casts on the very idea of trying to formulate such clear-cut theories of human evolution. Whatever the outcome, 1470 gives man longer to grow up. It is the latest of a series of discoveries in East Africa that have progressively extended further back, into the past, the time allotted to the evolution of human qualities. For Leakey, that is the chief satisfaction, for man is after all a very special animal:

> We're getting quite close to that extraordinary moment in time in Africa when animal investment in brains, hands, and sociability, began to produce something of a new quality – the human ploy, so remarkably effective and powerful that it has altered the very rules of the life game.

Sociable hunters

Living in groups has been an important theme in the evolution of primates and especially of humans. Great

Richard Leakey at work on a disinterred fossil elephant that is more than three million years old. Remains of animals other than man-like ones are very important for building up an impression of the living environment in which man evolved.

play has been made of comparisons between man and apes in social behaviour, to provide a biological gloss on human nature. But behaviour can evolve more rapidly than limbs. Man is unique among the primates in that his social behaviour evolved to suit practical requirements of hunting elusive or dangerous animals for food. A less naïve sociological biology is growing up which sees closer parallels for human social organisations in animals more distantly related but following a communal carnivorous life similar to that of early man: hunting dogs and hyenas, wolf packs and prides of lions.

The wild dogs of Africa are nomadic animals feeding off the gazelles, wildebeest and zebra of the plains. They have usually been detested and persecuted by man; only recently have students of animal behaviour begun to explore the subtleties of their entirely legitimate way of life. The dogs live in groups ranging from a mating pair of adults to about thirty animals, adults and young, but the characteristic way of life is hunting in packs, and the smaller groups probably represent outcasts from the packs. The hunting is very systematic with the animals assembling each day for the hunt, morning and evening; on moonlit nights they will hunt as well. First there is a ritual of greeting, which reinforces and synchronises the animals' social behaviour. Then they move off in single file towards the hunting grounds, where they may spread out, the better to find their prey. Acting in cohesion, the dogs can pass right through a crowd of game without any of them being distracted by this food near at hand, in order to reach a different quarry beyond.

An obvious advantage of communal life, which apes fail to exploit but wild dogs do, is the division of labour. Some adults remain behind to guard the pups while the others go hunting, and those returning from the hunt regurgitate food for adults and pups alike. A dog cannot resist the begging gesture of a pup rubbing its nose along the dog's mouth; indeed if another adult makes the gesture by accident during a ritual greeting, the animal automatically vomits. The wild dogs can breed all the year round, but there is a peak of births around the end of March. For the next couple of months, they remain in one place, guarding the den and the pups; except for this they lay no special claim to territory, although they will advertise their presence to other dogs by 'scent-marking' urination.

Lions are somewhat more 'territorial' in their habits, but the ranges of different groups overlap. The pride on which the lion's social organisation centres is typically a group of about half a dozen adult females occupying a particular territory, with two or three males living with them. As male cubs mature they leave their mother's pride and wander off singly or in small groups until such time as they may attach themselves to another group of females, if need be by killing or driving out the males in residence. Conversely, the males attached to a pride will defend their rights against intruders, but otherwise they rely on the lionesses to supply their needs. They will wait until the females have killed an animal and then barge in and help themselves to what they want – though they are then more generous than the lionesses in allowing the cubs to feed.

A squad of lionesses walking steadily in open line abreast towards their prey is perhaps the sight most strangely man-like in any animal behaviour. The advantages of group hunting are plain enough. While the dogs can kill zebras, lions acting in concert can tackle large animals including buffaloes and giraffes, which are more than a match for a solitary cat. And a single lion running into a herd of grazing animals will fail to catch anything, nine times out of ten; several lions

A pride of lions. Man differs in social behaviour from the apes, as these lions differ from other carnivores that hunt alone.

hunting co-operatively will succeed far more often, especially at night, among thickets, or when approaching the prey against the wind. Sometimes lions will lie in ambush; on other occasions they will appear openly and drive animals into a cul-de-sac.

These descriptions of wild dogs and lions are primarily those of George Schaller of the New York Zoological Society, who spent three years watching the animals of the Serengeti of Tanzania. Schaller comments:

> Man and his precursors . . . have been widely roaming scavengers and hunters for perhaps two million years, a way of life that has diverged so drastically from the nonhuman primates that similarities in the social systems of the two may well be accidental . . . . Knowledge of carnivore behaviour, as well as of primate behaviour, can provide the spectrum of possibilities open to the early hominids.

Dead ends

The old makes way for the new. That great carnivore, the cave bear, lumbered around the forests of Europe until just before the last retreat of the ice, when it became extinct. The new animal that displaced it was man, not in his modern guise with bullets and fences and spray guns but as a hunting animal pursuing the way of life for which he evolved. And although men hunted the cave bear sometimes, that was not the cause of its extinction. It was a matter of man as another powerful animal, with a similar mixed diet of meat and plants, competing with the cave bear for limited resources of food and shelter. The cause of extinction looks like hunger rather than direct action of the cave bear's enemies.

A disproportionate number of cubs were dying in the caves towards the end. Björn Kurtén of the University of Helsinki is able to trace the life and death of the cave bears in such detail, because the bears often died

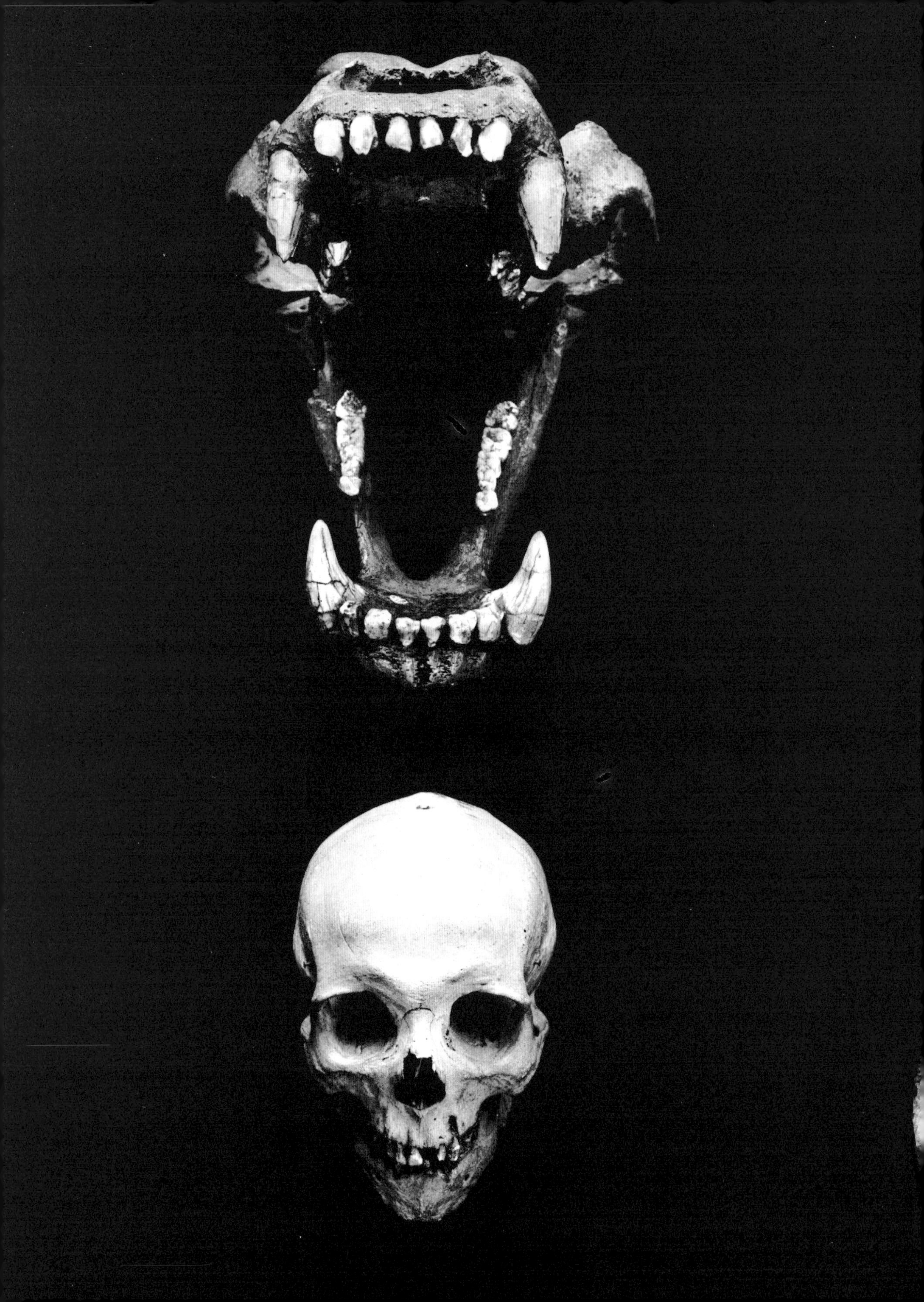

The jaws of a cave bear, with a human skull for comparison. If man hastened the extinction of the cave bear it was probably because of competition in the same way of life, rather than a consequence of hunting.

while hibernating in their caves. Fossil teeth of bears accumulated in vast numbers in a cave at Odessa in Southern Russia 27,000 years ago. Working with these teeth, Kurtén can sometimes diagnose the cause of death in individual cases; for example, malformed teeth that did not mesh properly with the opposing teeth would make it difficult for the bear to chew enough meat to store food before the winter hibernation. As one of Europe's best-known fossil-hunters, Kurtén says firmly that modern research on fossils must concern itself with individuals and with populations, seeking a thorough understanding of the ways of life of species. To talk loosely about whole groups of species as if they were the same things is old-fashioned, and bears are a case in point.

They started with wolf-like animals which, from the onset of the ice ages, about two million years ago, rapidly gave rise to several species of bears, down to the seal-hunting polar bear that first appeared near London during the last ice age 100,000 years ago. Each species had different ways of life and preferred different environments. But while the cave bear sought out its caves, and the forest-dwelling brown bear dug his own den, there were overlaps and the brown bear was to some extent in competition with the bigger cave bear. In England the brown bear even took to living in caves and seems to have kept the cave bear out of the area. Cave bears and brown bears were both very quick to change size, evolving to be larger in cold conditions, during advances in the ice, and becoming smaller again when the ice retreated. So conclusions about the fate of 'the bears', and why the brown bear survives and the cave bear does not, can come only from close attention to different populations within each species.

A simple score sheet of the appearances and disappearances of species can, though, reveal the overall rate of evolution. From wider-ranging studies of the history of mammals, Kurtén has been able to document an acceleration in evolution with the onset of the ice ages. Up to about a million years ago, mammal species were surviving, on average, for more than four million years before new species replaced them; since then many species have been enduring for less than a million years. But even that is not quite as fast a rate of turnover of species as occurred at the start of the age of mammals, seventy million years ago.

For calling the odds on survival in each phase of evolution, Kurtén uses 'half-life' – the period required for half of the species alive at a particular time to become extinct. The turnover rate varies from one group of mammals to another; the table below gives examples,

	70 to 60 million years ago	25 to 1 million years ago	250,000 to 10,000 years ago
Carnivores (e.g. cats, bears)	400,000	1,400,000	610,000
Primates (e.g. monkeys, apes)	500,000	1,600,000	230,000
Insectivores (e.g. shrews, moles)	400,000	2,400,000	490,000
All mammals	440,000	1,500,000	540,000

'Half-life', in years, for some groups of mammals at different times (after Kurtén). Low figures mean a fast rate of evolution. The mean longevity of species is almost three times the half-life.

including the particularly rapid evolution among the primates during the past million years. Numbers like Kurtén's may be needed for understanding evolutionary processes like those that caused the extinction of all the dinosaurs.

The rate of extinction is usually similar to the rate of emergence of new species. But extinctions exceed replacement when two isolated territories are united, whether because a stream dries up or because continents collide. The combined area can carry only a few more species than either area on its own. Conversely, new species come in faster than old ones are extinguished, when territory is fragmented.

All too little is known, though, about why particular species or groups of species die out. New species may be in some respects 'better' in the prevailing environment, but speculations about the genetic deterioration of the dying species turn out to be wide of the mark. For example, it was said that animals living a long time in a stable environment would narrow down the choices of alternative genes available in their populations, thus forfeiting all capacity for evolving if circumstances changed. This simply does not happen.

Animals living on the ocean floor, at depths of more than 1000 metres, experience as constant an environment as any on Earth, yet they possess just as much genetic variability as species living in shallow water on land. Again, some famous animals, often called 'living fossils', have persisted for tens or hundreds of millions of years with practically no change in form: the coelacanth fish, the spider-like horseshoe crab, and a little sea shell, lingula, are examples. Yet their living representatives show just as many alternatives in their genes as human beings do, or the newest Drosophila species out of Hawaii. 'Living fossils' may be unchanging but this is not by evolutionary neglect; the virtues of their designs are tested and reconfirmed at every new generation.

But mammals would not be able to reinvent gills in a hurry, if the world were flooded. All animals commit themselves to a more or less specialised way of life, from which they cannot back-track quickly if the environment changes drastically. Animal designs that are inherently versatile, robust and unpretentious are the ones with the greatest staying power. But the environment can always change more rapidly than the species can evolve and all species are probably, in that sense, a little obsolescent.

Successors for man?

If we survived a million years before following the cave bear into oblivion, we should still be doing better than most primate species in the present geological era. As we are living among H-bombs and a multiplicity of crises of our making, such a figure may even seem wildly optimistic – some might think a hundred years to be nearer the mark. Conceivably our easy ways of rationalising cruelty and folly could set at nought all our capacity for organisation, our technologies, our growing understanding of nature, our powerful brains and a stubborn will to live. The way some people talk about Doomsday suggests divine retribution for our overweening pride.

Supposing we did face extinction, I doubt whether even the fondest lover of animals would be happy to brood about the immediate replacements for man. Of course, the idea of 'dominance of the planet' is peculiarly human; tyrannosaurs or lions can have had no such awareness of a cosmic status in the living community. But there is an ecological sense in which animals that feed on animals that feed on animals that feed on plants – the 'top carnivores' – can exert a quite disproportionate influence on the entire ecosystem. If this is the criterion, species such as the bass and the eagles are our natural successors.

These animals lack, of course, other characteristics that we associate with dominance, particularly large numbers. By this test, among mammals, the rodents are quite strong contenders. But if we add coherent organisation as a requirement, the ants are the obvious candidates. Although the terminology of the biologists for 'worker bees', 'soldier ants' and the 'queen' imply analogies to human hierarchies, the social insects represent an entirely different strategy of organisation. Essentially brainless individuals operate very effectively like little automata and with perfect self-sacrifice for the well-being of the colony. It is often an ugly sight.

Would we prefer a brainier successor? For sheer mass of brain tissue, the whales far surpass even ourselves. But they have three great deficiencies. Whales engage in a charming sort of singing, but it seems not to begin to approximate to human powers of language. Without hands, their manipulative ability is small – but as that is a safeguard against technological imprudence, some might count it an asset. Can the same be said for the lack, in the whale's brain, of those 'frontal lobes' which give man his capacity for planning, for hope and for remorse, and for social interactions of the highest sensitivity?

Casting around for honourable successors to man, we must probably settle for the macaques and baboons. These are the adaptable and sociable monkeys of whom John Napier, a noted authority on the primates, has said:

> Were it not for the coming of man, macaques and baboons might well have been the dominant form of animal life in all the temperate regions of the Old World. . . . If man is President of the animal world, then the macaque is certainly the Vice-President.

Then we hear of Japanese monkeys discovering how to separate grain from grit by dropping the mixture in water and collecting the floating grain. But this is where we came in and the primates, investing in brains, hands and sociability, may not be easily bankrupt.

Happy the hare at morning, for she cannot read
The Hunter's waking thoughts, lucky the leaf
Unable to predict the fall, lucky indeed
The rampant suffering suffocating jelly
Burgeoning in pools, lapping the grits of the desert,
But what shall man do, who can whistle tunes by heart,
Knows to the bar when death shall cut him short like
the cry of the shearwater,
What can he do but defend himself from his knowledge?

W. H. Auden

Stitching genes together. The electron microscope shows small rings which are the normal genetic material (DNA) of the SV 40 virus, linear strands of the viral DNA that have been cut open, and large rings consisting of viral DNA which have been caused to join up with the DNA of a bacterial gene. The virus then becomes a means of transporting genes into cells.

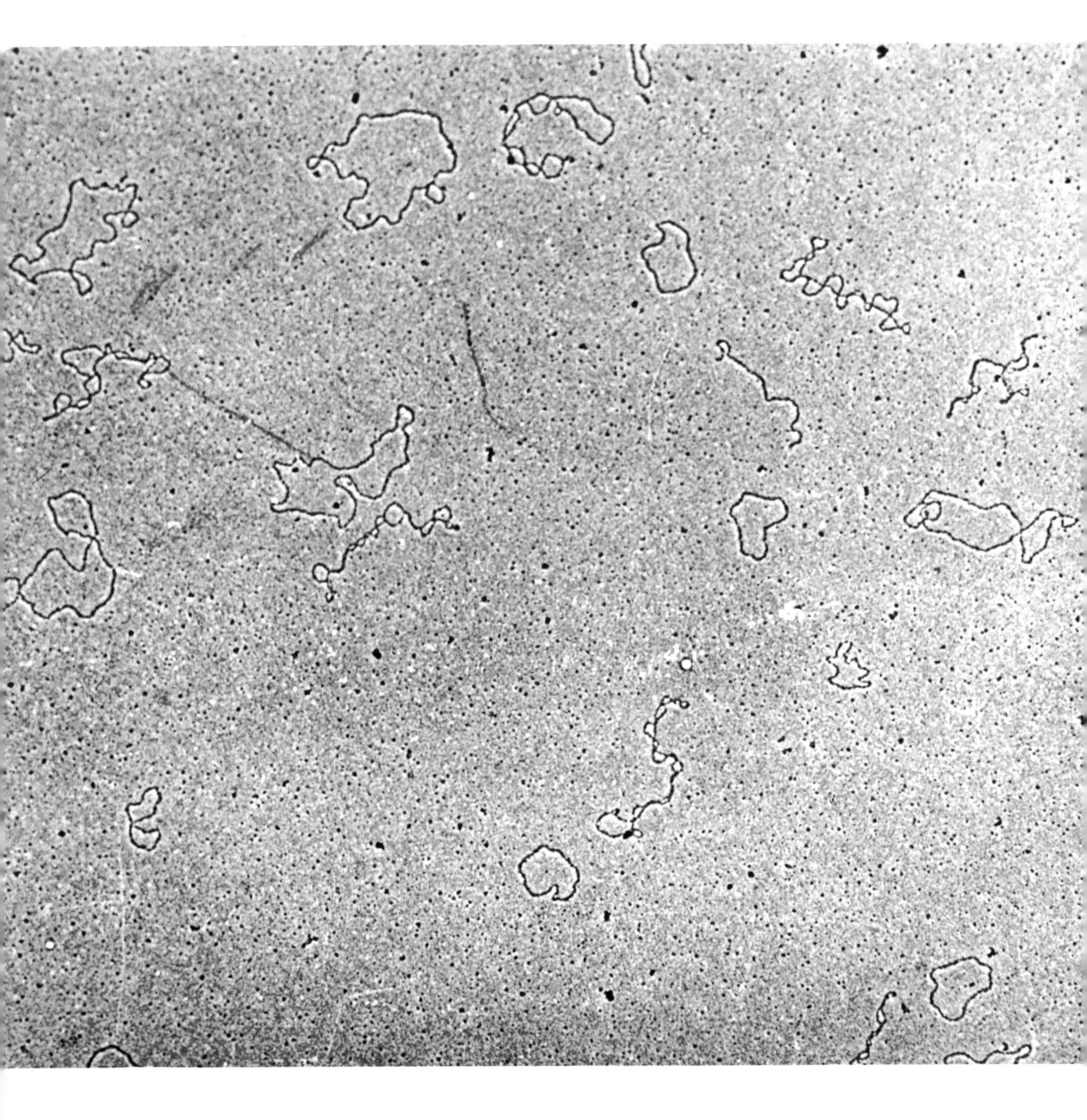

Postscript: A Game of Skill

> Our goal is to develop a method by which new, functionally defined segments of genetic information can be introduced into mammalian cells.

A group of biochemists at Stanford University in California announced, in these words, the development of a technique which is currently the outstanding illustration of the growing powers that humans have over the fundamental machinery of life. It uses viruses, which are in essence simply pieces of genetic material capable of invading living cells to secure their reproduction. By putting a chosen gene into a suitable virus, the virus becomes a truck for transporting the gene into the heart of the hereditary system. Biologists have speculated for a long time about the use of viruses for this purpose.

But Paul Berg, the leader of the Stanford group, objects to the label 'genetic engineering' being put on this work. He insists that his aim is an experimental technique for obtaining a better understanding of how genes work; our ignorance about the details of human heredity is far too great for us to think of bringing the genetic manipulation of man out of the realm of science fiction into medical practice. Berg worries, too, about the dangers of accidentally creating new virus diseases as a result of genetic manipulation and is critical of other scientists who are more carefree about experiments of this kind.

The virus SV40 has the ability to form a stable association between its genes and those of the cells that it invades. It then becomes heritable; in other words the cells copy the virus along with their own genes, and transmit it to their own daughter cells. The virus lies low and its presence makes very little difference to the cells. Introducing a new gene, so that it will be incorporated along with the SV40 genes, involves a fairly elaborate chemical process developed by Berg's colleague David Jackson.

The chosen gene, selected from bacteria, is a length of DNA; its ends have to be extended chemically with short pieces of artificial DNA for attaching to the virus DNA. The latter consists of a ring, which has to be broken open. On the resulting free ends go lengths of matching DNA which will join up with the ends prepared on the extraneous gene. When quantities of virus DNA and genes, thus prepared, are mixed together, they combine and reform rings in which the genes and the SV40 DNA coexist. The Stanford experimenters are using the virus to transport a gene from gut bacteria, which codes for working molecules used in breaking down galactose, a sugar. At the time of writing the modified viruses are ready but they have not yet been inoculated into laboratory cultures of living cells.

To suggestions that the way is now open for using such techniques for treating genetic disease, Berg replies:

> At the moment we see no way of directing a new gene into a specific location on the chromosome. . . . And unless one has absolute assurance that he can introduce a gene and only the one he wants in precisely the position that he wants to put it, I would have grave reservations about trying to do anything with humans – and no less animals. So I think the people who are going around advocating gene replacement therapy as a way of curing genetic diseases and promising that we'll have a solution within, not twenty-five, not ten but five years, are in fact misleading the public and promising something that science, I don't believe, will deliver.

A moratorium of a hundred years on any attempt directly to manipulate genes in human beings has been suggested, and seems like a good idea. Our ignorance of the genes and how they work is still so great that anyone tampering with human genes would be like an illiterate offering to improve the works of Shakespeare. But if we think about evolution we think in thousands and millions of years and it is hard to avoid supposing that, at some time, genetic engineering in man will begin.

During this century there has been a lot of small-minded talk among biologists about directed evolution in man, under the tag 'eugenics', including even proposals that women should gratefully queue up at the supermarket for the sperm of famous men. The implicit and explicit ideas about superior types of human beings all too easily converts into racism and Hitler was the only serious experimenter in eugenics so far. Most frequently the discussion among biologists centres on intelligence, as though it were the only human virtue, culminating in the notorious 'talking-point' of William Shockley that people who do poorly in IQ tests should be paid not to breed. There is anxiety, too, about genetic decline, because we keep people alive who would otherwise die of genetic defects. Diabetics are a case in point; yet a reduction in sugar consumption could probably cut the incidence of diabetes much faster than it will grow by keeping diabetics alive. In other cases medical protection will encourage defensive but otherwise harmful genes – including the sickle-cell gene – to disappear.

More charitable is genetic counselling about gross genetic defects; these can be voluntarily avoided if potential parents are warned about the precise risks entailed in dangerously mutated genes that they carry. But this is just good medicine; it is beside the point for the future of man because it achieves in a kindly, direct fashion what natural selection would do anyway, in the course of a few unhappy generations of malformed humans. The only approaches to the further evolution of man that are likely to be both effective and acceptable are those of positive genetic engineering of the kind foreshadowed – whether they like it or not – by the experiments of Berg and Jackson. Meanwhile, officious fussing about the genes and mating preferences of one's fellow humans is not a noble posture for the lords of creation.

New responsibilities

The life game went on feverishly for more than 3200 million years, with no one around capable of appreciating it. I find that every bit as vertiginous as the thought of all those billions of light-years of empty space around our planet. But the essence of the life game was that the players did not know what was going on. Whatever strategy they evolved to cope with chance and their environments was itself a product of chance and their environments. It was thus that natural selection achieved the lineaments of purpose out of purposeless events and made more complex organs and species, fitted as if by design to their roles in life. But organs and species could sometimes transcend their initial functions. A fin could become a limb, a limb could become a wing or the hand of a craftsman. A nervous system could become an articulate, inquisitive brain capable of mirth and remorse. And of purpose: with that paradoxical result the purposeless old game comes to an end.

With us a completely new life game begins – a game of skill, more like chess. For better or worse future evolution on Earth is in our hands. In detail, of course, in wild species and even in our own persons, the rituals of mutation and natural selection will continue. But a builder pays little enough heed to the atoms in his bricks. Whether we interfere actively with natural selection in wildlife, or simply set up enclaves where it can continue in its own way, or even revert to our bad habits of treating wildlife as a nuisance that must keep out of our way or perish – whatever it is, it will be our decision. Knowing so much about the life game, we cannot escape the responsibility.

The transition from the old game to the new is gradual. Ours is perhaps the very first generation to see how it is changing, thanks to a rather precise grasp of where we stand in space and time and our initial understanding of the mechanics of life. But the transition began when hunting men capable of naming other species appeared; it accelerated when the deliberate breeding of animals and plants started about 10,000 years ago; the discoveries of Darwin, Pasteur and Mendel each represented a big step towards terminating the old game. Our knowledge is growing apace and future generations will see the new game much more clearly than we do. But the crucial change was moral rather than technical. Despite the sentimentalists who say we should conform to nature, man long ago rejected the callousness of the old life game. Today we very properly shrink from applying to ourselves even the more calculated selective game of the breeders. For man himself only painless methods of directing evolution will do.

But thoughts for our own future must take full account of the facts of life. We have pressing biological needs that keep us dependent on the living and non-living resources of the environment, but issues of that kind are closer to economics and agronomy than to the evolutionary predicament of all other species. The most important lessons of modern evolutionary biology are (1) the uniqueness of every individual, (2) the immense possibilities genetically latent in every group of individuals, and (3) the falsehood of any notion of genetic perfection. We shall presumably never try to eliminate randomness from the genes, and here the controversy about the molecular heresy of Motoo Kimura (Chapter 3) is relevant.

The New Darwinists' theory adopts a curiously bivalent position about radical and conservative processes in evolution. In dealing with whole organisms they successfully disposed of the mutationists' belief in the ideal type to which individuals were supposed to approximate. The notion that replaced it, that of highly variable individuals in a slowly evolving population, represents a splendid new style of thought for the human mind. Yet at the molecular level the selectionists still regard individual genes as ideal types. A dash of neutral mutations, if confirmed, should strengthen 'population thinking' at its core.

Our growing knowledge of genetics and evolution should in principle enable us to keep the human flame alive indefinitely, so long as the Sun burns steadily (several thousand million years). The chief danger is that loss or suppression of our animal instinct for survival might let us decide that self-conscious life is a nasty practical joke of nature's; then we might die of discouragement. Otherwise, over the geological timescale that we can now contemplate, present legitimate misgivings about genetic manipulation of ourselves will be allayed by vastly enlarged knowledge. Then there will be new species of human beings designed by their predecessors.

New organisms

Long before that happens we shall have applied genetic engineering to farm animals and plants – thereby enlarging the already formidable powers of the breeder. Increasingly, too, new species will be domesticated, including some selected and adapted for industrial tasks at present allocated to machines.

Bacteria and other organisms that concentrate rare metals from their environment may replace the miners of non-ferrous metals and represent a microbial equivalent of the goose that laid the golden egg. The manufacture of other materials, greatly extending our present likings for silk, wool and wood, will be allocated to specially developed organisms. And presumably microbial goats will tend to solve our present problems of waste disposal.

We shall probably also take securely in hand the course of evolution in wild species of animals and plants which do not figure directly in the human economy – not merely protecting but perhaps greatly enlarging the areas set aside for wildlife. If, within those protected areas, man's creative urge is to promote an increase in the number of wild species, it is already fairly clear what has to be done. The first need is for action to counter the accidental transport of plants and animals between different parts of the Earth; our ships, aircraft and canals have been undoing the division of the world into separated land masses that followed the break-up of the last supercontinent and promoted a proliferation of species. The second step is to adopt a policy of Hawaiification of the protected areas – that is to say, artificially to divide them in such a way as to prevent genes from flowing between one enclave and another.

The isolated populations that resulted would evolve along different paths and create biological novelties in a quasi-natural way. The diversity of wildlife could be fostered too by inventing special environments: for example, by feeding warm water from power stations to artificial reefs, or constructing huge greenhouses to create semi-tropical conditions in the colder latitudes. The complete substitution of one kind of wild plant cover for another, or the deliberate introduction of alien species into particular enclaves will be bolder tactics open to us.

Pollinating insects, earthworms and soil microbes lie at a borderline between wild and domesticated species, as organisms on which farmers and game wardens depend. Fish, birds and other animals hunted for sport or food also fall into this intermediate category, and no doubt the ecology of trout streams, for example, will be artificially modified to promote the best evolution, fishing and eating. But at present it is among the soil microbes that the greatest opportunity seems to lie. Farmers add artificial fertilisers to the soil because of a shortage of the bacteria which naturally 'fix' the nitrogen of the air and make it available to growing plants; large installations and expenditures of energy now go into fixing nitrogen by chemical engineering. The working molecules of the bacteria which accomplish this capture of nitrogen with far less effort are being identified, together with the genes that make them. Already, by induced mating between different species of bacteria, the capacity to fix nitrogen has been transferred to species previously lacking it. The technical means seem near to hand for an unprecedented fertilisation of the whole Earth – after careful thought about the dangers of overdoing it.

New people

When, eventually, human beings turn their genetic skills upon themselves, what may they attempt to do? There are certain rather elementary maladaptations of man to his densely populated and largely sedentary life of today. One is his status as a carnivore. It is biochemically necessary for man to eat meat or other animal products; this is sometimes disputed by reference to the vegans, the strictest vegetarians of India, but it turns out that they unwittingly obtain essential nutrients from the maggots that infest their food. Man would be much less reliant on animal food if he could be equipped with enzyme systems capable of making vitamin B_{12} and the essential amino acids; our need for vitamin C from fruit and vegetables, a rather rare defect among animals, would also be ended if we could make it ourselves.

Genetic cosmetics may be irresistible. If men and women of the distant future contrive to be uniformly beautiful, that may mark the end of an important aspect of natural selection in man, because a person's appearance will no longer be much of a guide to physical and mental fitness. It is hard to predict what the skin colour of choice will be. Other variations on the present human theme would be to make us very big and strong or, more plausibly, small and compact so that our requirements for food and living space would be correspondingly reduced. In the latter case, problems might arise about brain size. We could have large heads on small bodies, but either babies would have to be born even more immaturely than they are already or the female pelvis would have to be re-engineered. But most of the possibilities already mentioned are comparatively trivial in the sense that we should still recognise these modified people as members of our own species. Imagination begins to falter when it comes to envisaging entirely new species of men.

Australopithecus, reflecting with half a brain on his world of volcanoes and vultures and pint-sized women, would not have had the wit to envisage a being such as *Homo sapiens sapiens*. And although we tag our own species in that self-satisfied way, we may be just as lame in speculating about species with faculties exceeding our own. When people stop to think about this, as they rarely do, they tend to vote for more of the same. They look to an amplification of the mental powers that most plainly set us apart from other animals. Suggestions have ranged from babies born talking perfect English to massive sessile brains inhabiting concrete turrets instead of bodies. But this preoccupation with brains when thinking about the very long-term future may be as much a mark of narrow-mindedness as it is in the short-term concerns of the eugenicists. Almost everyone who does consider the long-term future is an intellectual living in one of the present-day communities that evaluate brainpower like silver ingots and impertinently assign IQ scores to the unfathomable genius of children.

The gods visualised in other eras were not notable eggheads. Belligerent farmers revered physical strength and magical powers over nature, especially over the germinating seed; sexuality counted for more than sagacity. High craftsmanship, the artist's touch or the poet's tongue certainly rely upon mental powers but not in the sense that IQ-testers normally mean. Only in the last few years have brain researchers become fully aware of the vast amount of brain capacity allotted to non-verbal skills, which our educational systems have grievously neglected. Another large volume of the brain, the frontal lobes, is concerned among other things with sensitivity of social relationships; there is ample provision too for control of the muscles, whether in the delicate work of the craftsman or musician or the more strenuous movements of the athlete or dancer. As there is almost unlimited opportunity for cultivating our existing mental powers more thoroughly and in more balanced ways, we need not think of asking for better brains than those which sufficed for our hunting forefathers.

Above all other considerations, man is a social animal and the effect of amplification of brain power that comes from the coherent action of a human group is far more relevant to our foreseeable ways of life than a lot of individual mental giants would be. In this very sense, we should need to know far more than we do about the interplay between the intellectual, emotional and housekeeping functions of the brain, before we could even begin to guess how to expand it without creating inhuman monsters. By the time that knowledge exists, man may have succeeded in organising a society free from the anxieties about survival or economic and military competition of the kinds that have encouraged the recent emphasis on particular mental skills. If our descendants decided to act, they would be as likely to attend to the cerebral lobes concerned with music as those dealing with arithmetic.

The radical possibilities for self-directed human evolution may well have little to do with brains. Here again the myth-dreams of our forefathers, about gods with wings or fishtails or horses' bodies, may be to the point. Of these, wings may be the most enchanting idea. But if our descendants preferred to live under water they could equip themselves with gills. They may fancy fur that will let them face the elements anywhere without the need for special clothing. Or they may want to live for ever, providing themselves with brains capable of self-repair and bodies as fit for self-renewal as the hydra's. Other possibilities exist without analogues in the natural world; for example, a thoroughgoing symbiosis between human beings and mechanical and electronic devices.

Our species has always observed and marvelled at living things. In a few more generations our understanding of how we come to be here may be as thorough as the imperfect record of the distant past will ever allow. By then, too, our descendants will know more exactly how our bodies and brains work. The revolution in attitudes that Darwin started will be complete when the interplay between the genetic messages from our animal ancestors and our own purposeful, upward-reaching talents is fully grasped. Then the criminal follies that flow from treating people as if they were either fallen angels or expendable chemical machines will have abated. Only then will our species be in the right frame of mind to contemplate the next 1000 million years, and wonder what game it should play.

Index

Italic entries indicate pages where the topic is referred to only in the caption to an illustration.